GW00992338

THE CITY

LONDON UNITED TRAMWAYS

HISTORY OF THE CITY

LONDON UNITED TRAMWAYS

A history, 1894–1933

GEOFFREY WILSON

LONDON AND NEW YORK

First published in 1971

This edition published in 2007
Routledge
2 Park Square, Milton Park, Abingdon, Oxon, OX14 4RN

Routledge is an imprint of Taylor & Francis Group, an informa business

Transferred to Digital Printing 2007

© 1971 Routledge

All rights reserved. No part of this book may be reprinted or reproduced or utilized in any form or by any electronic, mechanical, or other means, now known or hereafter invented, including photocopying and recording, or in any information storage or retrieval system, without permission in writing from the publishers.

The publishers have made every effort to contact authors and copyright holders of the works reprinted in the *The City* series. This has not been possible in every case, however, and we would welcome correspondence from those individuals or organisations we have been unable to trace.

These reprints are taken from original copies of each book. In many cases the condition of these originals is not perfect. The publisher has gone to great lengths to ensure the quality of these reprints, but wishes to point out that certain characteristics of the original copies will, of necessity, be apparent in reprints thereof.

British Library Cataloguing in Publication Data
A CIP catalogue record for this book
is available from the British Library

London United Tramways
ISBN10: 0-415-41753-8 (volume)
ISBN10: 0-415-41933-6 (subset)
ISBN10: 0-415-41318-4 (set)

ISBN13: 978-0-415-41753-2 (volume)
ISBN13: 978-0-415-41933-8 (subset)
ISBN13: 978-0-415-41318-3 (set)

Routledge Library Editions: The City

LONDON UNITED TRAMWAYS
A History – 1894 to 1933

Geoffrey Wilson

ILLUSTRATED

LONDON
George Allen & Unwin Ltd
RUSKIN HOUSE MUSEUM STREET

FIRST PUBLISHED IN 1971

This book is copyright under the Berne Convention. All rights reserved. Apart from any fair dealing for the purpose of private study, research, criticism or review, as permitted under the Copyright Act, 1956, no part of this publication may be reproduced, stored in a retrieval system, or transmitted, in any form or by any means, electronic, electrical, chemical, mechanical, optical, photocopying, recording or otherwise, without the prior permission of the copyright owner. Enquiries should be addressed to the Publishers.

© *George Allen & Unwin Ltd, 1971*

ISBN 0 04 388001 0

PRINTED IN GREAT BRITAIN
in 11 point Plantin type (2 point leaded)
BY W & J MACKAY & CO LTD, CHATHAM

LUET, LUET, let me see.
Of course, 'Let there be light'.
                    ANON

*We honour our chairman,
we serve the public and we
trust in God.*
        JAMES CLIFTON ROBINSON

Acknowledgments

I should like to record my gratitude to the following for supplying information and material, without which this book would not have been possible:

Messrs J. Allen and G. Baddeley; Reverend Peter W. Boulding; Mr Anthony Bull; Miss Florence D. Comerton; Messrs R. Elliott; C. Hamilton Ellis; G. Gundry; Alan A. Jackson; J. Joyce; C. F. Klapper; Addison H. Laflin, Jr (Bay Area Electric Railroaders Association); Charles E. Lee; A. W. McCall; and Walter McGrath; Professor George W. Hilton (University of California); Mr Thomas O Grady; the late Mr Eoin O Mahony, KM; Messrs Alan T. Newman; G. Rogers; L. J. Ross (Electric Railroaders Association Inc., New York); R. Tustin; and Cyril Smeeton (Hon. Publications Officer, Tramway and Light Railway Society); Mrs O. Warrell; Mr Bryan Woodriff; A. Randow Ltd.; The Chamberlain Court (City of London); Press Office, London Transport; General Manager, Bradford City Transport; Publicity Officer, Bristol Omnibus Co. Ltd; Royal Society of Arts; Borough Surveyor, London Borough of Merton; Institution of Electrical Engineers; British Transport Archives; Greater London Record Office (Middlesex Records); City Librarians, Cork City Library and City of Westminster Public Libraries; Borough Librarians of the London boroughs of Ealing, Hammersmith, Hillingdon, Hounslow, Kingston, Merton, and Richmond, and their helpful staffs at the main and district libraries (particular thanks are due to Miss W. M. Heard, Librarian in charge of Reference and Information Services, Chiswick District Library); Borough Librarian, Middlesbrough.

I am most grateful to Miss Carol Dennett for relieving me of much typing in the final stages and to my son Terry for helping me to check the typescript. My thanks go to Mr J. Braidley for producing the map and to Mr Mark Shearman for help with the photographs.

Any errors and omissions are mine. I shall be grateful to all readers who point them out.

GEOFFREY WILSON
October 1969

Merton Park,
London, SW 19.

Contents

PROLOGUE	*page*	17
1. Enter Robinson		29
2. The Opening Shots		35
3. Battle Grounds		44
4. Electrics at Last		57
5. Junkets – And Hard Bargains		62
6. New Ground – And 'Underground'		78
7. Mainly Robinson		90
8. Hampton Court and Uxbridge		95
9. Into Surrey		108
10. Rounding Off		115
11. Chill Winds – And New Brooms		125
12. A 'Tramway King' Passes		132
13. LUT, MET and BET		137
14. Wartime Readjustment		145
15. Spencer Makes His Mark		150
16. Mild Revival		157
17. Modernization		164
18. Finale		170
EPILOGUE		179
APPENDIX		
I Routes and Service Numbers, by *A. W. McCall*		181
II Fares and Tickets, by *A. W. McCall*		200
III The Fleet		226
IV Rules and Regulations		229
BIBLIOGRAPHY		233
INDEX		234

Illustrations

1. One-horse single-deck car
2. Reckenzaun's battery car
3. Acton High Street   *facing* 16
4. Pair-horse car in Chiswick
5. Pair-horse car outside the new Acton depot
6. Kew Green terminus
7. Kew – Richmond horse car   17
8. London's first public electric tramway opens
9. Inaugural cars arrive at Ealing Town Hall
10. Chiswick car-shed ready for the inaugural banquet   32
11. No. 101, 'flagship' of type X
12. Lower deck of the first LUT electric cars
13. Upper deck of the first LUT electric cars   *between* 32–3
14. Type X No. 141
15. Conductor alters destination indicator of type X No. 133
16. Type X No. 126   32–3
17. Young's Corner, Chiswick
18. Cars passing outside LUT offices in Chiswick High Road
19. Type Z No. 6   *facing* 33
20. Type X No. 117
21. Type W No. 184
22. Nos 4 (type Z) and 182 (type W)   48
23. Cars 7 and 19 of type Z
24. At Hounslow ('Bell') terminus
25. No. 61 of type Z (later Y)   *between* 48–9
26. Shepherds Bush in May 1903
27. Holiday crowds at Shepherds Bush   48–9
28. Type X No. 143
29. Type X No. 125
30. Nos 335 (type T) and 148 (type X)   *facing* 49
31. Type X No. 132
32. No. 112 on Ealing Common
33. Type X No. 138   64
34. Type Z (later Y) No. 78

List of Illustrations

35.	Type W No. 210	
36.	Type W No. 223	*between* 64–5
37.	Type T No. 336	
38.	Hillingdon on the newly-opened Uxbridge route	
39.	Type Z (later Y) No. 15 at Uxbridge terminus	64–5
40.	Type W No. 212 at Richmond Bridge terminus	
41.	Cars in King Street, Twickenham	
42.	Type W (later U) No. 295 on left and type W No. 248	*facing* 65
43–47	Specimen tickets	*between* 80–1
48.	LCC cheap midday ordinary and workman tickets issued by LUT on route 89	96–7
49.	Cars on inspection pit tracks in Fulwell depot	
50.	Type W No. 249 inaugurates Teddington service in 1902	
51.	Type W (later U) No. 212	96–7
52.	Bottleneck in Hampton Hill	
53.	Type W (later U) No. 266	
54.	Type W No. 173	*facing* 97
55.	Sir James Clifton Robinson	
56.	LUT 'Serenader' in Garrick Villa grounds	
57.	LUT band at Garrick Villa	112
58.	Holiday crowds by the river at Hampton	
59.	Guests of Sir Clifton Robinson at Garrick Villa	
60.	Sir Clifton and Lady Robinson at Garrick Villa	*between* 112–13
61.	Type W No. 196	
62.	In Broad Street, Teddington	
63.	Widening by Peg Woffington's cottage, Teddington	112–13
64.	Disruption in Eden Street, Kingston	
65.	Triangular junction at Hampton Wick	
66.	Hampton Court terminus, with type W car No. 296	*facing* 113
67.	The Mayor of Kingston takes type T No. 320 across Kingston Bridge	
68.	Bedecked cars cross Kingston Bridge	
69.	The Mayor of Kingston drives No. 302	128
70.	Bruce drives No. 321 on BOT inspection	
71.	Type T No. 330, heading for Kingston Hill	
72.	Type T No. 309 at Hampton Wick in 1906	*between* 128–9
73.	Surbiton, with No. 168 of type W	
74.	Victoria Road, Surbiton	
75.	Type T No. 327	128–9
76.	Car on King's Road shuttle	

List of Illustrations

77.	Type W No. 162	
78.	Dittons terminus with T No. 312	*facing* 129
79. 80.	} Type T No. 213 in Kingston Road, New Malden	
81.	Type W No. 43	144
82.	Type W (later U) No. 287	
83.	Type W (later U) No. 269	
84.	Type W No. 200	*between* 144–5
85.	Type W (later U) cars Nos 272 and 274	
86.	Type W No. 156	
87.	Type W No. 260	144–5
88.	Type W No. 180 on Summerstown service	
89.	Tooting-bound car at Wandle Bridge, Merton	
90.	Type W No. 263 with crew	*facing* 145
91.	'Electric cars stop here if required'	
92.	'Electric cars stop here'	
93.	'Cars stop here on Sundays during hours of divine service'	
94.	Parade of cars at Fulwell depot	160
95.	Type U (ex-W) No. 296 and type W No. 168 with LCC cars at Tooting	
96.	LCC car at Wimbledon Hill terminus	
97.	LUT and LCC cars at Kew Bridge	*between* 160–1
98.	Contretemps in Kingston	
99.	Type U (ex-W) No. 300	
100.	Type W No. 240	160–1
101.	LUT conductress, Mrs E. Seal	
102.	Type Y (ex-Z) No. 48	
103.	Barber truck on No. 52	
104.	Cars Nos 141 and 142 converted to 'flood cars'	*facing* 161
105.	Special car for private hire	
106.	Pullman-type saloon of one of the private hire cars	
107.	Type S2 No. 342	176
108.	No. 341, the first PAYE car	
109.	Moving route indicator inside PAYE car	
110.	Vestibule of PAYE car No. 342	
111.	Type S2 No. 344	*between* 176–7
112.	Type T No. 324	
113.	Top deck of reconditioned No. 307	
114.	The Hounslow Club and Institute (LUT) wives' and children's outing	176–7

List of Illustrations

115.	'Poppy' No. 350	
116.	Type T No. 320	
117.	No. 261 at Shepherds Bush in 1928	*facing* 177
118.	Type T No. 335	
119.	No. 211 which became type WT in 1928	
120.	No. 396, last of the 'Felthams'	192
121.	'Feltham' type No. 353	
122.	Saloon of a 'Feltham' car	
123.	'Feltham' No. 356 at Uxbridge	*between* 192–3
124.	4-wheel ticket van No. 4 (later 004)	
125.	Water car in original condition	
126.	Water car in final condition	
127.	Bogie stores van No. 005	192–3
128.	Type W No. 254 and trolleybuses	
129.	Trolleybuses at Twickenham	
130.	Trolleybus No. 33 at Teddington	*facing* 192
131.	Modern view of the former Chiswick power station	
132.	Curved staircase and gallery in the former Chiswick power station	
133.	Wellington Road entrance to Fulwell Depot with trolleybuses	208
134.	Trolleybus in Wimbledon Hill Road	
135.	Trolleybus No. 43 in original condition	
136.	74-seater centre-entrance experimental trolleybus	209

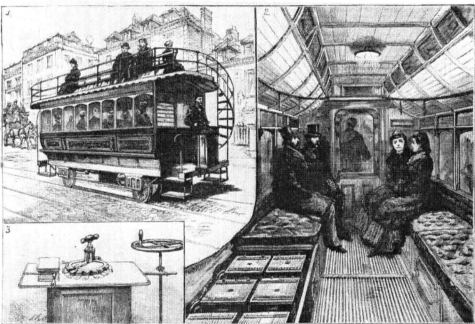

Top: West Metropolitan Tramways one-horse single-deck car on Shepherds Bush–Young's Corner Service. (*O. J. Morris Collection*). *Centre:* Reckenzaun's battery car tried out at Kew Bridge in 1883. (*London Transport*). *Bottom:* Acton High Street in LUT horse-car days. (*Courtesy Acton Library*)

Left: LUT pair-horse car on Hammersmith–Kew Bridge service. Note the 'Bristol' style of the number on the dash. *Right:* Bedecked for Queen Victoria's Jubilee in 1897 is this LUT pair-horse car outside the new Acton depot.

The Kew Green terminus of the Kew–Richmond horse tramway. (*Courtesy Richmond Library*)

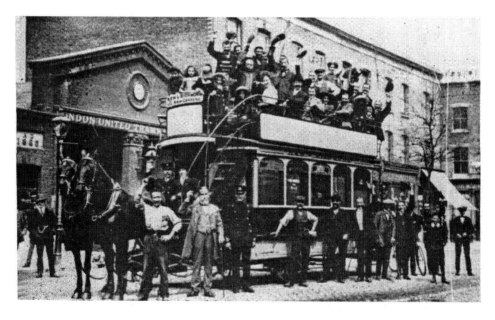

Kew–Richmond horse car outside Richmond depot, possibly on its final run in 1912.

Prologue

'Will you walk or have you time to take a tram?' So ran a local joke in the West London of the early 1890s. The butt was the ramshackle West Metropolitan Tramways, a system which after a few years of operation had failed to make a financial success of running its four horse-car routes.

The West Metropolitan had started off in good style in 1881, taking over a Shepherds Bush–Acton horse-car line from an earlier operator and quickly adding new routes, one along Goldhawk Road to Chiswick, another from Chiswick to Kew Bridge, yet another south of the river, from Kew Green to Richmond, and finally a link between Hammersmith and Chiswick.

It had a goodly heritage. The routes, along main roads, including two of the main western highways, had a sound traffic potential. Competition from horse buses and often roundabout and slow steam-worked railways was hardly vigorous, and the districts traversed were growing rapidly.

Its constructional burst, subsequent fruitless attempts to reduce operating costs by finding a workable alternative to horse traction, and a series of ambitious projects to extend both into central London and to the west seemed to sap any energy that the management may once have possessed. By 1893 when its long-suffering debenture holders successfully appointed a receiver, the only hope seemed to lie in selling the whole concern for as good a scrap value as could be obtained – or in finding a saviour.

Nucleus of the West Metropolitan, the Shepherds Bush–Acton tramway, opened in two sections in 1874 and 1878, was one of the first permanent tramways to be laid in London, where tramway construction was lagging behind that of the provinces. London might have led the way in Britain in 1857 if the newly-formed London General Omnibus Company – founded two years earlier under French auspices – had been successful in promoting a tramway between Notting Hill and the Bank by way of Marylebone and the City Road, with a branch from Kings Cross to Fleet Street.

Street tramways were an American invention which sprang from the circumstance that some early American railways were laid down main streets of towns. It was a logical step to lay a line for local use only. The first authenticated street railway began in Manhattan in 1832 as a downtown prolongation of a proposed steam railway, the New York and Harlem.

Paris was the first city in Europe to adopt tramways. Alphonse Loubat opened a demonstration line in 1853 and began public service in 1855, hence the Paris-based 'General' company's initial interest in running both trams and buses in London.

In 1859 John Curtis began operating a 'rail-bus' service in Liverpool, running his vehicles over the Mersey dock lines, but it was across the Mersey, in Birkenhead, that Britain's first true tramway was opened, with great éclat, on August 30, 1860.

It ran between Woodside and Birkenhead Park and it was the first British venture of eccentric American George Francis Train, of whom more will be said later. Train had tried without success to gain Parliamentary powers for tramways in this country. He then patented his special rail system and began to build lines by agreement with local authorities.

Although the LGOC had proposed to use a flush-type rail, so as not to inconvenience other road users, Train adopted an L-shaped step rail with the top of the vertical part laid slightly proud of the road surface. His method did not recommend itself to cabmen, carriage drivers and carters and his flamboyant manner cannot have endeared him to local bigwigs.

The line, though relaid with grooved rails, was short-lived. But it made an intense impression on a Birkenhead youngster who was destined to make his mark on the urban transport scene.

Train laid step-rail lines in West Derby (Liverpool) and London in 1861. They were equally ephemeral. So also was a line in Darlington, and only a change to grooved rail after Train had quit the British stage saved his Hanley–Burslem tramway from sharing the early demise of its fellows.

Train's first plan for London comprised a large loop line, from Finchley Road to Baker Street, Wigmore Street, Regent Street, Oxford Street, Portman Square, Gloucester Place and back. He was unlucky in that there was a railway boom at the time, so that tramway schemes failed to attract support, and was foolish in persisting in his step rail.

He had to settle in London for three short, disconnected lines, all opened in 1861. The first was along Bayswater Road, between Marble Arch and Porchester Terrace, and was opened on March 23, 1861, in the presence of a great crowd of celebrities, including Charles Dickens and Douglas Jerrold. It was followed on April 15th by a line along Victoria Street between Victoria Station and Westminster Abbey, and on August 15th by one from the south side of Westminster Bridge to Kennington Gate.

Train's lines were laid on the understanding that they were to be removed at short notice if authority required. It soon did. The Bayswater line closed in the September, the Victoria Street line in March 1862 and the Kennington line in June 1862.

The first two routes were never again to see tramways but the third would later become part of the South London trunk tramway system.

Sustained tramway development in this country should perhaps be reckoned from the inauguration of the Liverpool Tramway Company's service in November 1869. The company was the first tramway undertaking to gain an Act of Parliament.

The Tramways Act of 1870 began as a sincere attempt by the Board of Trade to regularize the procedure for promoting and laying down tramways. Had it passed in the form intended, urban transport might have developed differently. But the Bill had a rough passage through Parliament and it emerged laden with irksome provisions. Board of Trade certificates which had needed no Parliamentary sanction were rejected in favour of Provisional Orders, which required such approval. Frontagers, that is, owners and occupiers of property along the proposed route, gained the right to object in general, and of veto where the line was to run within 9 feet 6 inches of the kerb for a length of 30 feet and a third of their number opposed. The sanction of street authorities was still necessary, but – a truly British compromise – promoters gained the right to overcome local authority vetoes if their line was to run through several districts and was supported by authorities for at least two-thirds of its length!

Even so, the benevolence of the Metropolitan Board of Works augured well for the prospects of tramway promoters in London and several tramway entrepreneurs hoped to extend their activities to central London, not merely the suburbs. But such powerful opposition was mustered that all projects for tramways in the City and West End were postponed for consideration by a joint select committee of both Houses.

The effect of the findings of this committee – which also recommended that the procedure for Private Bills should be brought into line with that for Provisional Orders – was effectively to seal off central London from tramways, with the result that, for good or ill, London was never to gain a truly comprehensive tram network. The psychological effects were as severe as the physical. In London, tramways came to be regarded as all very well for the suburbs, particularly the less fashionable, but not the thing for the City or West End.

The first permanent tramways in the inner suburbs all opened in 1870: the Brixton–Kennington section of the Metropolitan Street Tramways (May 2nd); the Whitechapel–Bow section of the North Metropolitan Tramways (May 9th); and the New Cross–Blackheath Hill section of the Pimlico, Peckham and Greenwich Tramways (December 13th).

Although Train's Bayswater venture had been a bad choice of route, as it served only a well-to-do district with a carriage-owning population, there were good possibilities farther west, based on the traffic centre of Shepherds Bush, which was served by the Metropolitan and West London Railways.

On May 12, 1870 the Southall Ealing & Shepherds Bush Tram-Railway (later Tramway) Company was incorporated to build and work a line from Uxbridge Road Station (Shepherds Bush) along the Uxbridge Road to Acton, Ealing, Hanwell and the 'Red Lion' at Norwood (Southall). In 1872, before construction had begun, its engineer, George Billinton, tried unsuccessfully to promote an extension from Southall to Uxbridge.

In the same year Richmond Vestry considered the construction of a tramway along Kew Road. It was the start of an equivocal attitude to tramways by the Richmond authority which was to last for forty years.

On December 16, 1873, Reid Brothers, of City Road, began work on the first section of the SE & SB, between Uxbridge Road Station and the 'Princess Victoria' (Askew Crescent), Acton Vale, a distance of 1 mile 10 chains. It was opened on June 1, 1874. The cost of £5,500 a mile was said to be lower than that of any other metropolitan tramway yet built. Even so, the line was not a financial success and it ceased working on February 20, 1875, when revenue was stated to be £40 a week and working costs £27.

Reid Brothers took control and the Board of Trade sanctioned reopening on September 21st. When the Board of Trade heard the owners' application in July to extend through Acton, it was said that 1,000 passengers were using the tramway daily and that some £19,000 of the authorized capital of £25,000 had been spent.

The owners gained a Provisional Order in 1876 to extend 1 mile 5

Prologue

chains to a point 25 yards west of 'Priory Road' (Acton Lane). The extension was opened on February 18, 1878.

Meanwhile there had been abortive attempts, in 1874 and 1875, to build other lines in connection from Shepherds Bush to both Notting Hill and Kensington.

Although the LGOC had long since given up any aspirations to operate both trams and buses, it was quite willing to supply horses to tramway operators. Accordingly from February 18, 1878, it horsed the Acton tramway by agreement with Charles Courtney Cramp, to whom the line had been leased for £600 a year.

By this time there was agitation for tramways in the Isleworth district. In an editorial in its December 14, 1878, issue, the *Richmond and Twickenham Times* said: '. . . it is a notoriously inconvenient fact that there are few districts so badly off for the means of locomotion as that lying between the Middlesex side of Richmond Bridge and Brentford. . . . Compared with the antique, doleful and jolting "bus", the tram car is almost a palace on wheels and in point of comfort it is far beyond the average of South-Western Ry. carriages. . . .' The knowledge that a road was used by trams generally put other drivers on their guard, 'and thus in actual practice it has been found that on thoroughfares where there is this supposed need for special caution, and therefore its exercise, accidents least frequently occur'.

The engineer to the Brentford, Isleworth & Twickenham Tramways, then promoting a Bill, took up the newspaper's theme, saying that experience in London showed that even in narrow streets tramways aided traffic regulation.

The BI & T Bill was for 5 miles 39 chains of line between Kew Bridge, Brentford, Isleworth, St Margarets and the Middlesex side of Richmond Bridge. Frontagers in both Brentford and Isleworth were in favour but the wealthier residents of 'East Twickenham' were indignant. When the Bill came before the Commons the solicitor for Lady Chichester said that it would be 'intolerable to have the tagrag and bobtail disgorged before her ladyship's lodge'.

The scheme was shorn of its St Margarets section and passed as a line between Brentford Bridge and North Street, Isleworth.

In 1878 there were proposals for tramways between Hammersmith Broadway and both Kew Bridge and Barnes. In 1879 the Brentford & Isleworth company sought to extend from its authorized line at Busch Corner, Isleworth, to both Hounslow Heath and Twickenham and lay a

Prologue

branch from Hounslow ('Bell') for some distance along the Bath Road towards Cranford.

New vigour seemed about to be injected into West London tramway development, sadly languishing, by the incorporation on August 12, 1881, of the West Metropolitan Tramways Co. Ltd, with a capital of £100,000, to acquire the Shepherds Bush–Acton line and build extensions.

The new concern began energetically. In March 1882 it took over the Acton line and by Act of August 10, 1882, was reincorporated as a statutory company under the same title.

Powers were gained at the same time for lines from Hammersmith Broadway through Turnham Green to the Middlesex side of Kew Bridge; from the Surrey side of that bridge, alongside Kew Gardens, to Richmond (Lower Mortlake Road); and from Shepherds Bush along Goldhawk Road to join the Hammersmith–Kew Bridge line at Young's Corner. (Young was a greengrocer whose shop stood on the corner of King Street and Goldhawk Road.)

Kew Bridge, predecessor of the present structure, was considered too narrow for a tramway, a fact which was profoundly to affect the course of tramway history in the area.

The routes were opened as follows:

Shepherds Bush–Young's Corner.	March 18, 1882
Youngs Corner–Kew Bridge ('Star and Garter')	December 16, 1882
Kew Bridge (south side)–Richmond (Lower Mortlake Road)	April 17, 1883
Hammersmith Broadway–Young's Corner	July 14, 1883

The Kew Bridge (north side) section was soon the scene of an interesting experiment in battery traction. Anthony Reckenzaun, an Austrian settled in England, tried out a converted horse car in which fifty cells had been placed under the seats in the lower saloon to supply power to a motor driving one of the two axles. Electric lamps and bells were also installed.

It was said that the batteries could operate the car, with a full load of forty-six passengers, for seven hours. The top speed achieved was 6 m.p.h.

Although the experiment was claimed as a success, it apparently lasted only a day, March 10, 1883. According to *The Graphic* the car had

Prologue

to be helped up a rise by horses after a connecting band had failed.

Throughout the 1880s there were abortive efforts to promote new tramways in West London. Some were by the West Metropolitan, which in 1884, for example, sought powers for 8 miles 13 chains of double track and 6 miles 23 chains of single track in Acton, Hammersmith and Southall. In 1885 the Brentford & District Tramways promoted a Bill to incorporate and build tramways from Kew Bridge to Isleworth and Twickenham, and to Hounslow Barracks, totalling about eight miles of line.

The B & I company's track in Brentford and Isleworth had apparently become disused by this time. The Hounslow Local Board resolved in 1889 to lift it between Brentford Bridge and Busch Corner and between Isleworth Union and Mill Plat. At the same time the West Metropolitan was examining the possibilities of extension into Brentford.

There was a scheme by the Acton & Hammersmith Tramways in 1887 to incorporate a company to build lines between Askew Arms and Hammersmith via Askew and Paddenswick roads, and from Hammersmith along Hammersmith Road as far as Avonmore Road, West Kensington, 2 miles 39 chains in all.

In view of later developments it was unfortunate that the West Metropolitan could not follow up quite a Parliamentary triumph in 1889, when, optimistically, it was authorized to extend along the Uxbridge Road to Ealing and Hanwell; from Shepherds Bush via Morland, Latimer, Lancaster, Cornwall, Westbourne Park and Porchester roads to Harrow Road; from Askew Arms to Glenthorne Road, Hammersmith; and lay new track between Uxbridge Road Station and Acton Vale.

The new lines, totalling 5 miles 67 chains of double and 5 miles 24 chains of single line, were to be worked by cable or other system. The company also gained power to raise additional capital of £187,500, though two years before it had tried to get its capital reduced.

This ambitious programme seems almost like bravado when the company was struggling hard enough to keep its existing system going. It can be explained only by the parallel of the Metropolitan and Metropolitan District Railways and their thrust ever deeper into the suburbs in order to bring fresh traffic to the expensive Inner Circle.

Batteries had not provided the answer to the quest for mechanical traction and, like other tramways at this time which had ruled out steam haulage, the West Metropolitan began to seek some form of propulsion in which the power was transmitted from a central source.

Prologue

It is necessary here to digress briefly to consider what progress had been made in applying electricity from such a source. On May 16, 1881, Werner von Siemens – who had shown a small electric locomotive at the Berlin Exhibition of 1879 – began public service on a 1½-mile line in Berlin which used the running rails as conductors. The obvious disadvantages soon led to the provision of a third rail, as adopted by Britain's first electric lines, the Giant's Causeway Portrush & Bush Valley Railway & Tramway and Magnus Volk's railway on Brighton beach – both opened in 1883 – and the Bessbrook & Newry of 1885.

But a third rail, unless protected in some way, was out of the question for a street railway. Both in Germany and America there were experiments with a troller, a small collector running on top of an overhead wire and flexibly connected to the car. The American ex-naval lieutenant Frank J. Sprague perfected the under-running trolley arm and wheel method. His system opened at Richmond, Virginia, on February 2, 1888, was an immediate success and heralded the remarkable spread of electric tramways during the next twenty years.

In more conservative Britain, aesthetic and other reasons seemed at first likely to preclude any general adoption of overhead trolley tramways. On September 29, 1885 a conduit electric tramway, designed by Michael Holroyd Smith, was opened along the seafront at Blackpool, ancestor of Britain's last remaining urban tramway today. The slot system for cable traction had already been perfected in America and had lately been applied in London on Highgate Hill. But the vestries in the West Metropolitan's area, as well as the Metropolitan Board of Works, objected at that time to a slot in the road for electric traction.

A way out seemed to be offered by the Lineff closed-conduit system, with which the West Metropolitan conducted trials between 1888 and 1890, using a 200-yard section of specially-equipped track on the west side of Chiswick Depot.

Between the running rails were laid 3-foot sections of ordinary rail whose top was flush with the road surface. Below this rail but not in contact with it was a strip of flexible hoop iron laid on an electrical conductor. When a car came over a given section of track, a powerful magnet under the car lifted up a 3-foot rail section and allowed the motors to be energized. When the car passed the rail dropped back to its original position, the process being repeated all along the line.

Although high hopes were entertained of this device – the West Metropolitan hoped to reduce working costs by 50 per cent by using it –

Prologue

the Lineff system joined the great ranks of highly ingenious but over-complex systems that were never put into practice. Even to function moderately well, it would certainly have required meticulous laying and maintenance.

An abortive Bill of 1890 was that of the West London Tramways Company, which sought incorporation and powers to build lines from Acton to Hanwell and Hammersmith, and in Kensington, Paddington and Fulham. In the following Session the promoters asked for power to reincorporate the company and build tramways specifically in Uxbridge Road, Askew Road, Paddenswick Road, The Grove, Hammersmith Road, Shepherds Bush Road, Richmond Road (Shepherds Bush), Norland Road, Lancaster Road and Tavistock Road, and to have running powers over part of the West Metropolitan.

In 1891 the West Metropolitan hopefully informed the public in the *West London Advertiser* of its plans for complete relaying and electrification. It also proposed to revive its 1889 powers to extend from Acton to Hanwell and from Hammersmith to Addison Road.

Speaking at the company meeting on March 1, 1892, E. H. Bayley, the chairman, expressed confidence in the future, despite 'one of the most trying years experienced', and looked forward to the adoption of electric traction at no distant date. The company had spent £5,742 on providing for double track between Hammersmith and Kew Bridge – a work completed in 1893 – but the vestries concerned agreed to maintain the track at their cost for the first year.

The company was nothing if not a trier. Although by 1893 it was pretty decrepit, it asked for powers not only to build new lines and substitute new for existing lines but also to use steam, electric or other mechanical power, divide the undertaking into two or more parts, authorize Hammersmith Vestry to lend £20,000 towards the cost of building certain lines, raise fresh capital and again extend the time for authorized lines.

It was a bold gasp but a dying one. On June 13, 1894, the receiver whom the debenture holders had appointed the year before offered the undertaking for sale as a going concern. The assets were listed as:

Chiswick Depot (just off Chiswick High Road, between the present Merton and Ennismore Avenues), with stabling for 140–170 horses, a 3-storey granary, fodder stores, sheds for 20 cars, seven cottages arranged as 14 dwellings for workmen, a yard and an exercise ground Shepherds Bush Depot (between Shepherds Bush Road and Wells

Road) with stabling for 61 horses and a yard for seven pair-horse cars
Richmond Depot (Kew Road) with stabling for 30 horses and a covered yard for 6 cars
34 pair-horse and single-horse cars
15 new pair-horse cars, being supplied by G. F. Milnes, Birkenhead

The sum of £750,000 had been spent on the system and a revenue of £25,000 derived.

The condition of the West Metropolitan, or at least of the Acton line, which seems to have been the worst part, may be judged from an article in the September 1894 issue of *Railway World*. 'The local authorities have maintained with some show of reason that when their officials returned from an early morning stroll, laden with debris of the permanent way, in the shape of pieces of rail 3 or 4 feet long, they were justified in holding that the tramways in their streets were unquestionably a nuisance and, as such, should be removed.'

The writer said that to compare the West Metropolitan to a modern tramway was like comparing a London Chatham & Dover third class coach to the comparable accommodation of the Scotch Express. He went on: 'Only a personal inspection of the museum of antiquities now in the yard of the Chiswick depot can give an adequate impression of the extraordinary condition to which both rolling stock and permanent way had been allowed to fall. Some of the cars would serve admirably as a 'pre-historic peep', while the rail sections, ranging from the stringer and Kincaid to the more recent girder pattern, deserve the study of the Tramways Institute.'

It is not surprising that there was no mad rush to secure such a prize. In fact no bids were made at the auction on June 13, 1894. But behind the scenes, in Bristol, there was great activity. On August 6, August Krauss, tramway contractor of Bristol and director of the City of York Tramways, bought the West Metropolitan assets for £30,000 through the Court of Chancery. The *Railway World* of August 1894 understood that Krauss would spend £25,000 on doubling in Acton and Hammersmith and would put new cars on the Acton and Kew Road routes. The track, it reported, had fallen into a dangerous state, but because of the purchase an inquiry into the condition of the Acton line had been suspended.

Krauss bought the West Metropolitan on behalf of a new company, the London United Tramways Co. Ltd, of which he was a signatory. The company had been registered in Bristol, with an address at Clare

Prologue

Street House, and had been incorporated on July 19, with commendable speed in view of its aim. Its capital was £25,000 of ordinary, £25,000 of preference and £32,000 of debenture shares.

George and Samuel White, secretary and assistant secretary of the Bristol Tramways Company, were the backers and the Imperial Tramways Company – of which more anon – had a large holding. The other signatories were:

William Butler, J.P., Chairman of Bristol Tramways Company
Joseph Kincaid, M.Inst.C.E., Chairman of City of York Tramways Company
H. G. Gardner, Deputy Chairman of City of Gloucester Tramways Company
Alderman Bartlett, Chairman, Bartlett & Son Ltd
Hugh G. Doggett, solicitor, Bristol Tramways Company, and Coroner for Bristol
James Clifton Robinson, A.Inst.C.E., Managing Director, Imperial Tramways Company.

Within two years the purchasers would transform the old West Met. into one of the best-built, best-equipped and best-managed horse tramways in the kingdom, and pay an 8 per cent dividend into the bargain. They would amply justify the shrewd assessment of potentialities made by Robinson after he had inspected what other men might have dismissed as a hopelessly unprofitable field for speculation.

The new proprietors soon showed their colours. They announced that they were ready to find all the money needed for rejuvenation. They would scrap all cars beyond repair and modernize those in reasonable condition. New 46-seater cars designed by Robinson would be ordered from Milnes and a five-minute headway service would be provided on some sections. (The modernized stock included the Lineff trial car shorn of its gear and renovated for service on the Kew Road route.)

The London United – the name suggested ambitions beyond inner West London – gained its first Act on July 6, 1895. It confirmed Provisional Orders for new lines, mostly double track, in substitution for parts of existing lines in Hammersmith, a new line in the Grove, Hammersmith, and an extension (3 furlongs 2.6 chains of single and 1 furlong 5.58 chains of double track) from the Acton terminus at Birkbeck Road to Acton Hill ('The White Hart'), and a new depot at Acton Hill in the

Prologue

angle of Uxbridge Road and Gunnersbury Lane to replace that at Shepherds Bush.

No time was lost in getting to work, The Shepherds Bush–Acton line was doubled throughout. New tracks were laid with hardwood between the metals and granite paving for 18 inches outside the outer rails. The West Metropolitan's 1887 Order as to motive power was repealed to allow cars to be moved by animal, electric or any mechanical power other than steam.

Robinson, as Managing Director of the LUT, declared: 'Later on, when the question of conversion to mechanical traction is taken up, as it must be, we shall require to spend more than the outlay at present contemplated but we shall be only too pleased to do so if the local authorities desire it.' He added that the signatories to the company's articles of association really were the owners, all men with long practical experience in tramway work.

Drivers, conductors and inspectors were put into liveries of uniform type and more men were engaged. New horses were brought and new cars, designed by Robinson, were ordered or built by the LUT at Chiswick, where the depot was enlarged.

Chapter 1

Enter Robinson

Two significant facts about the new-formed London United Tramways Company were that it was Bristol-born and that James Clifton Robinson begat it. The facts are interwoven. To understand how we must consider Robinson's already remarkable career.

By 1894 James Clifton Robinson, at the age of forty-six, had established a formidable reputation in the tramway world. He thought big and by a combination of charm and pertinacity, combined with a masterly grasp of potentialities, he gained valuable backing.

He was born on December 31, 1848 – a future tramway Stephenson born in the year of George Stephenson's death – son of William Robinson 'of no profession', of Birkenhead. Even his birthplace was auspicious. As we have seen, it was in Birkenhead, on August 30, 1860, that Train laid down Britain's first street tramway.

Robinson, a boy of twelve, was fascinated by Train's tramcar. How much formal education he had received we do not know. But he was a bright lad and it was the age of self-help. He did not rest until he had become Train's office boy.

In 1866 Robinson went with Train to the United States, where, in New York, he gained valuable experience in building, operating and managing street railways. He was back in England in 1871 as assistant to the American firm of Fisher & Parrish, railway and tramway contractors, on the construction of tramways in London, Liverpool and Dublin.

He was appointed general manager and superintendent of works of the newly-built Cork tramways in 1873. The six double-deck cars bore his name on the side panels. He was on the way up. One day, on the steps of the Imperial Hotel, Cork, he was introduced to Mary Edith Martin, daughter of Richard Martin of Blackrock, Co. Cork. They were married in 1874.

Robinson's wife was a beautiful blonde, who kept her looks almost until she died. As her husband was a fine-looking man they must have made a handsome pair. She was imperious, used to having her way, had excellent taste and proved an admirable hostess.

Robinson's first English appointment came in 1875, as first general manager of the Bristol Tramways Company, another new company, formed by Joseph Kincaid.

He made a great success of things in Bristol, where he first became associated with George (later Sir George) White, whose success story almost matched his own. He and White made a great team. It was the triumph of the Bristol tramways, coupled with the financial acumen of George White and his brother Samuel, and their talents and Robinson's, which would ensure remarkable expansion later on.

The Robinson's only son, Clifton, was born in Bristol in 1880.

In 1882 Robinson became general manager and secretary of the Edinburgh Street Tramways Company. He reorganized and extended the system and introduced Sir James Gowans's continuous steel girder rail.

He began to ponder the possibilities of mechanical traction on tramways. Possibly as a result of seeing the inauguration of the Giant's Causeway Portrush & Bush Valley line in 1883 – he indicated the possibilities of electricity as a motive power for tramways in a paper on cable traction which he gave to the Royal Scottish Society of Arts in that year.

Robinson's paper attracted such notice that he was called on to organize and operate the Highgate Hill cable line, opened in May 1884 by the Steep Grade Tramway & Works Co. Ltd and the first cable tramway in Europe. Between 1884 and 1886 he managed the parent concern, the Hallidie Patent Cable Tramways Corporation Ltd (A. S. Hallidie had introduced his cable system to San Francisco in 1873. His associate E. S. Eppelsheimer brought it to Europe).

Possibly through Robinson connections, the Edinburgh Northern Tramways, a company formed in 1884, built for cable traction that year and in 1887 two lines which the Edinburgh Street Tramways Company had not constructed. It made an agreement with the Hallidie Corporation and opened the first line in 1888.

While managing the corporation Robinson appeared before the Parliamentary committees considering the City of London & Southwark Subway Bill, put forward in 1884.

The Highgate Hill line in its original form was no great success and the corporation went into liquidation in 1888. But by then Robinson had crossed the Atlantic once more.

Enter Robinson

He returned to the States at the invitation of Los Angeles. On July 8, 1887, the Los Angeles Cable Railway was incorporated to build a system of 60 miles, serving all the populated area of that rapidly booming city and extending to Pasadena and Santa Monica. Only 10½ miles were built but they served most of the city save the south-east and a hilly area where two earlier cable lines operated.

On October 31, 1888, the LACR promoters sold a majority interest to C. B. Holmes of the Chicago City Railway. Holmes organized the Pacific Railway Company, which took over the undertaking on September 9, 1889.

The new system was brought into use between June 8 and December 7, 1889. As at Highgate, the Eppelsheimer bottom-grip method – still surviving in San Francisco – was adopted. The system included three long viaducts to take the cable cars over steam railways. One, 1,535 feet long over the Southern Pacific RR, carried two tracks on a viaduct supported by single columns.

Holmes put in Robinson as general manager. In his *Los Angeles Railway Interurbans Special No. 11*, Ira L. Swett gives a colourful picture of Robinson at the time. 'If we are to place full credence in the reliable reporters of that period" the author writes, 'he was a flamboyant individual, given to eccentric and flashy maneuvers to bring himself to the attention of his less-colorful brethren. His chief delight was to drive at breakneck pace in his handsome buggy through downtown streets, cutting corners as closely as possible, all the while shouting greetings to business acquaintances as they craned their necks in a usually futile effort to pierce the cloud of dust that marked his passing.' Robinson had certainly learnt the trick of self-advertisement from his old chief Train and added some touches all his own.

As was the custom, Robinson received the honorary title of colonel. The 'colonel' and his lady must have presented quite a spectacle to Angelinos when they were taking the air. No doubt little of this was lost on young Clifton!

A deluge on the night on December 24, 1889, wiped out the Second Street Cable line. It also damaged the Pacific system, though not irreparably. The slots were filled with sand and debris and the power houses were flooded.

Robinson went here, there and everywhere, urging on the gangs to clear up the mess. On Christmas morning a business acquaintance twitted him about the absence of cars.

Robinson was touched on the raw. He bet a cigar the cars would be running again by 1 p.m. When that time came he ordered the engines restarted, although hours of work were still needed to restore the lines to safe working.

He won his cigar, but the cost was heavy. The abrasive sand and dirt were ground into cables, pulleys and machinery. The large sums necessary to repair the damage caused by Robinson's bravado hit the new concern hard. Robinson was asked to resign, if not fired. He had been accused of diverting flood water from the main power house to the basements of adjacent buildings, but was cleared.

Such a setback would have blasted the careers of many men. But not a Robinson. Northwards he went to San Francisco, where in 1890 he was practising as an electric railways consultant. He had taken with him to Los Angeles H. T. Jones, who had been the first gripman at Highgate. Jones, who had been appointed a division superintendent of the Pacific, went to San Francisco too and took a job as a conductor with the Market Street Cable Railway Company. By 1915 Jones had risen to become general superintendent of the United Railroads of San Francisco. He gained praise for his service during and after the 1906 earthquake and fire, and the strike that followed.

In 1889 the American Street Railway Association had appointed Robinson to report on mechanical traction. His investigations took him to Texas and Mexico and to Canada. The outcome was a notable paper, presented at Pittsburgh in 1891, which must have done much to wipe out any stain of the Los Angeles affair. He was cautious enough to conclude with the observation that 'no system of traction is capable of universal application'. All the same his thoughts were increasingly turned towards electric traction, in particular towards tramways operated on the overhead trolley system successfully inaugurated by Frank J. Sprague in Richmond, Virginia. Sprague's success began a boom in electric transit in the United States, though, as Robinson remarked years later, zeal often outran prudence.

He returned to England in 1891 to become managing director of the Imperial Tramways Co. Ltd. The company had been incorporated in 1878 to acquire several tramways, which under combined management might be made more efficient and profitable. They were a mixed bag, which included the Dublin Southern District, the Gloucester, Middlesbrough and Reading tramways, and the Corris Railway. By the time Robinson became associated, the Imperial had also acquired the Bristol

Top: Electricity triumphant. The official inauguration of London's first public electric tramway on 10 July 1901, with a caravan of new type X cars about to leave Shepherds Bush for Southall with guests of the LUT. (*London Transport.*) *Centre:* The inaugural cars arrive at Ealing Town Hall from Southall for the official opening by Lord Rothschild. (*Courtesy Ealing Library.*) *Bottom:* Chiswick car shed gay with flags and bunting for the inaugural banquet. Clifton Robinson (in frock coat) in centre front, with guests and LUT officials. (*London Transport*)

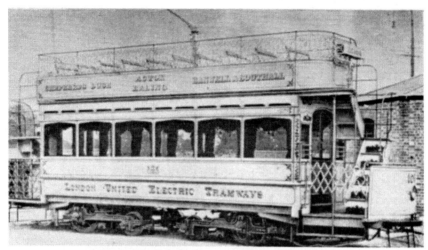

Top: No. 101, 'flagship' of type X. (*Courtesy R. B. Parr.*) *Centre:* Cushions, curtains and floormat originally graced the lower deck of the first LUT electric cars. (*Courtesy Alan A. Jackson.*) *Bottom:* Upper deck of the first LUT electric cars, showing throw-over type garden seats and iron grille work. (*Courtesy Alan A. Jackson.*)

Type X No. 141 on the Hammersmith loop in Beadon Road, outside the Met. & G.W. Hammersmith station. Note bracket arm poles, rare on the LUT. (*Courtesy Hammersmith Library*)

The conductor alters the destination indicator of type X No. 133 in Glenthorne Road on the Hammersmith loop. (*Courtesy Hammersmith Library*)

No. 126, type X, westbound in King Street, Hammersmith. Eastbound track turns left into Studland Street. (*Courtesy Hammersmith Library*)

Above: Young's Corner, Chiswick, with car from Shepherd's Bush (left) waiting to follow car from Hammersmith on right. (*Courtesy G. L. Gundry.*)
Below: Cars passing outside the LUT offices in Chiswick High Road. In foreground is type Z No. 6 on Kew Bridge–Shepherds Bush service. (*Courtesy R. B. Parr*)

Left: No. 6 of type Z stands outside the company offices at Chiswick on the depot approach track.

system, of which, of course, Robinson had intimate knowledge.

One of Robinson's first important assignments was to report on the desirability of electrifying the Bristol tramways. The growing district of Kingswood, east of the city, had long wanted better communication but the hills were considered to be too steep for horse trams and the potential traffic too light to justify cable traction. In 1891 the Bristol directors had obtained a Provisional Order to use electric traction on any extension from St George's to Kingswood.

When the Board of Trade determined regulations for electric traction early in 1894 the way to Kingswood seemed clear. Robinson's report was encouraging and was fully endorsed by George White, as managing director. Bristol Corporation and the St George's and Kingswood authorities being co-operative, no time was lost in securing parliamentary powers.

First, the Old Market Street – St George's line was relaid and extended to Kingswood to form a 3½-mile route. Joseph Kincaid designed new 67-pound rail, British Thomson-Houston contracted for the electrical equipment. Milnes of Birkenhead built twelve 4-wheel tramcars, prototypes of what were to be, in essence, the standard Bristol car throughout forty-five years of electrical operation!

By August 8, 1895, work was advanced enough for a horse car to be run to Kingswood to check the gauge. It carried Robinson, George White and C. Challenger, the traffic manager. Just twenty years before, the three had been associated in opening the city's first tramway.

The electrified line was opened to the public on October 14, 1895.

Although beaten by the Roundhay line at Leeds (1891), the South Staffordshire installation of 1893 – not to mention the Guernsey of 1892 – the successful inauguration and operation, and early extension of the Bristol system were important milestones in British electric traction history.

Meanwhile in Dublin Robinson was active in the electrification of the Dublin Southern, which was to be the first application in Britain of high-tension a.c. transmission to substations.

When the Imperial Co. appointed Robinson to manage the Dublin Southern that system was still in two sections. One 5-foot 3-inch gauge line linked Blackrock with Haddington Road, Dublin, whence cars continued over the Dublin United's metals to Lower Abbey Street. The other line, of 4-foot gauge, ran from Kingstown station to Dalkey. Linking the sections was the 5-foot 3-inch gauge line of the Blackrock & Kingstown Tramway Co. Ltd.

Robinson reviewed the DSD. It was resolved to buy the B & K and modernize the enlarged system. An Act of 1893 authorized conversion of the Kingstown-Dalkey line to broad gauge and electrification throughout. Powers to electrify were gained only after a great struggle, which braced Robinson for other contests ahead. At a dinner given to the staff in 1894 Robinson said that the company had had practically to fight the whole battle of electric traction for the three kingdoms. Any driver, he added, could handle an electric tram after a day's training. When the service was fully operating they could run a faster 5-minute service in place of a 15-minute service at 5 m.p.h. and employ ten times the number of men.

A successful trial run was made on November 30, 1895, from Shelbourne Road depot, Dublin, to Kingstown and back, with Robinson at the controls. As even with horse traction the revivified company was carrying almost 2,000,000 passangers a year between Dublin and Dalkey, great things were expected. Again, Milnes supplied the cars.

Robinson was one of the first tramway managers to fit out his men in smart uniforms. Those on the modernized DSD were dark blue with white piping. A peaked pillbox cap was worn.

Mr Alan T. Newman has recounted the exciting inauguration of electric services. Robinson was in London when at 4 p.m. on Friday May 15, 1896, he received Board of Trade certificates permitting public opening. He wired the staff to be ready next morning and caught the night Irish Mail. On the steamer he met the Rt Hon. Richard F. McCoy, the Lord Mayor of Dublin. Both men hurried from Kingstown to Dalkey, where the Lord Mayor performed the opening ceremony at 8 a.m.! Before the close of the service at 11.30 p.m., 25,000 people had ridden the electric cars.

Robinson was soon bent on extending the DSD in Dublin, if necessary in competition with the Dublin United, which was by this time considering electrification. The corporation was sympathetic but the opposition of the DUT was enough to throw out the DSD Bill. The DUT then moved into the attack and successfully induced sufficient DSD shareholders to sell their holdings and so give it control.

Robinson soon had far too much else on hand to be dismayed by events in Ireland. And at least it was his drive and vision which had brought to Dublin a system of electric traction which was to be for many years a model.

Chapter 2

The Opening Shots

The LUT took over 160 horses and 20 cars at Chiswick, 30 horses and 6 cars at Richmond, and 61 horses and 7 cars at Shepherd's Bush. Most cars were double-deck open-top vehicles with knifeboard seats on top. Some cars used later on the Richmond line were former conduit cars from Blackpool, where this form of operation lasted from 1885 to 1899. By 1898 there were 59 cars and more than 400 horses to run a vastly better service.

The 'Askew Arms' – Birkbeck Road section was discontinued for four months to enable the track to be relaid and wood blocks put down. In expectation of electrification Robinson shrewdly had the new rails electrically bonded.

Sanction in 1895 for the Acton Hill extension also conferred power to electrify throughout subject to the consent of the London County Council and local authorities. It was to prove one of Robinson's greatest tasks to secure general consent.

Even with horse traction, the LUT, like the rejuvenated DSD, was a remarkable improvement on what had gone before, as the *Acton & Chiswick Gazette* recognized in its issue of Saturday September 7, 1895: 'On Saturday last the new section of the Acton tramline from opposite the "Windmill"–the old terminus–to the top of Acton Hill was opened for traffic, and it was at once seen what an immense public improvement has been carried out by Mr J. Clifton Robinson and his co-directors of the London United Tramways, Limited. The new section is nearly three-quarters of a mile in length, and the whole of the line from Uxbridge-Road Station has been relaid with annealed rails weighing $92\frac{1}{2}$ pounds to the yard. These are said to be the heaviest rails used in any tramway in the world, and they are laid on a solid bed of concrete, the line being wood-paved from end to end. Every approved modern improvement has been introduced in relaying the line, and the work has been admirably carried out by Messrs G. Wimpey and Co., of The Grove, Hammersmith.

The whole of the cars running on the line are new, and built in the most up-to-date style by Messrs G. F. Milne and Co., the world-renowned car-builders, of Birkenhead. New harness has also been provided, and a splendid lot of young Irish horses have been introduced to work the cars. We understand that the company has acquired a large piece of land on Acton-Hill on which will be erected commodious stables and a suitable car depot. This work will be commenced forthwith. Workmen's cars now run every morning from Acton-Hill right through to Uxbridge-road Station, and on the "Chiswick" to Hammersmith Broadway at suitable times to catch the early workmen's trains. To celebrate the opening of the Acton extension, and in fulfilment of a promise made by Mr Clifton Robinson some time ago, the double fares hitherto charged on Sundays and Bank Holidays were abolished last Sunday, when the ordinary weekday fares were charged on the whole of the lines under the control of the company. It is needless to say that the old prejudice against the trams is rapidly fading away in consequence of the many important public improvements which are being introduced on the lines from time to time by the enterprising Managing Director, who seems determined to win success by deserving it.'

In December 1894 Robinson invited the members of the works committee of Hammersmith Vestry to see the overhead trolley system which had lately been inaugurated at Le Havre. It was a smart move. He could have taken them to Leeds, Walsall or Guernsey. But one imagines that the vestrymen were not averse to a trip to France, even in winter! On his return the vestry surveyor wrote a report wholly in favour of allowing the LUT to try overhead electrification. Robinsonian blandishment and advocacy had won the first round in West London. Many rounds were to follow.

At the time the *Railway World* wrote: 'While it is perhaps possible that a certain section of the London County Council may object to the introduction of electric traction on the ground that improved facilities for locomotion may tend to discourage the masculine citizen to absent himself from the domestic roof after sunset, there will probably be no serious objection and, as the company have already obtained Parliamentary powers for the electrical equipment of the lines, London will then have an example of electric tramways on the overhead wire system that will compare favourably with many admirable lines on the Continent.'

The LCC would indeed prove obstructive, but on technical rather than moral grounds!

The Opening Shots

Notwithstanding all the provincial welter, Robinson was active in London. By March 1898 he was campaigning vigorously to win over still hostile or lukewarm authorities. Ealing, which had always kept the West Met. at bay, now feared that overhead trolley tramways would lower property values.

Robinson was equal to that. He wired authorities in places with electric tramways: 'Impression created here that the introduction of overhead trolley electric tramways depreciates the value of property, and lowers the character of districts through which the cars run. It is also suggested that the electric lines are more dangerous, and, on scientific, sanitary and economic grounds, afford no greater advantages to a community than horse tramways. Kindly wire opinion on these points based upon your experiences.'

Replies from Blackpool, Bristol, Coventry, Douglas, Dover, Dublin, Guernsey, Hartlepool, Kingstown, Leeds, Walsall and Wednesbury all supported Robinson.

On Saturday, March 5, 1898, appeared the first issue of an illustrated weekly called the *Chiswick Electric Tramways News*, a penny journal published by William Hutchings of 143 Uxbridge Road, Ealing, and obviously Robinson-inspired if not Robinson-edited. A serial entitled 'The Chronicles of Chizeek' began: 'Now in ancient time in the land of Chizeek was a road called the Ramshackle Road . . .' The 'hero' of the piece was the good Cliftonides!

Robinson afterwards said that he had founded three newspapers for tramway propaganda and when their work was done had cheerfully extinguished them. Being an amateur newspaper proprietor and editor was, he considered, far harder than being a tramway manager.

By this time Robinson had taken a house at No. 345 Goldhawk Road and called it Pasadena. Obviously the Los Angeles affair did not weigh heavily!

Robinson was to fight three major pitched battles for the LUT – Ealing, Teddington and Kingston. By 1897 it would have been difficult to poll a hundred pro-tram votes in Ealing. Robinson sent out to every resident a pamphlet entitled 'The Case for the Trams' and followed up with a canvasser armed with a petition praying the council to consent to the LUT plan.

Even the pulpit was invoked by the anti-tram party, who predicted the ruin of the town if Robinson was successful. But a progressive party, with electric tramways as a main plank, was formed. Robinson discreetly retired – as a non-resident he could not properly intervene in local politics – but continued his missionary labours.

But the anti-tram party was returned by one vote. Robinson was patient. Eighteen months went by. Then the opposition suddenly collapsed. Before the next elections, trams in Ealing would be an accomplished fact.

The Light Railway Act had been passed on August 14, 1896. Although the Act was clearly intended to simplify the promotion of lines to serve rural districts at a time of agricultural depression, the preamble did not attempt to define a light railway.

Three permanent commissioners appointed by the President of the Board of Trade considered schemes. If they reported favourably the promoters gained their powers under an Order and were spared the expense of going to Parliament.

Robinson saw at once how the imprecise wording of the Act might be turned to advantage. Admittedly a Light Railway Order could not be sought if the proposed line lay wholly within the limits of a single local authority, and it might be opposed by a main-line railway on the grounds of competition. But on the other hand local authorities could not exercise their vexatious right of veto under the 1870 Act and the procedure was cheaper and quicker than a Bill. Tramways paid rates in full, but light railways paid only a quarter of general district rates. And if new light railway was but old tramway writ large, any blame lay with the draftsmen of Westminster.

In May 1898 the LUT deposited an ambitious Bill for new tramways and for the electrification of existing and future lines. The tramways comprised an extension from Kew Bridge through Brentford to Hounslow ('Bell') (3 miles 5 furlongs) and one from Acton Hill through Ealing to Hanwell (Brent Bridge) (3 miles 2 furlongs), and a line branching at Brentford Half Acre and traversing Boston and Lower Boston Roads to the Uxbridge Road, just east of Brent Bridge, Hanwell (2 mile 3 furlongs). Powers were also sought to build power stations on existing depot land at Acton and Chiswick.

The Boston Road line was partly an insurance in case the company could not gain powers for the contentious Ealing line.

Hanwell was not thought of as a terminus. A Light Railway Order was sought for an extension of 7 miles 2 furlongs along Uxbridge Road to Southall, Hayes and Uxbridge. It was the first extension of an existing or proposed tramway to be planned as a light railway. Another Order was

The Opening Shots

sought for a more unusual line, to run from the south side of Kew Bridge by way of Sandycombe Road, Manor Road and Queens Road to Richmond Hill, thence via Star and Garter Hill, Petersham and Ham, to Kingston (4 miles 7 furlongs), not only to tap the important Surrey administrative centre and market town, but also to attract the lucrative summer pleasure traffic to Richmond Park. A bridge was needed to avoid crossing the LSWR on the level in Manor Road.

The Uxbridge light railway scheme was generally welcomed by the places it proposed to serve. Uxbridge, served only by a Great Western Railways branch, was particularly enthusiastic. At the beginning of 1898 a petition in the town drew 6,000 signatures in favour. The council adopted the resolution and sent it to the LUT.

The Light Railway Commissioners heard the application at Southall on June 29, 1898. There was fierce opposition from the Great Western Railway, Middlesex County Council and some frontagers but not enough to deter the Commissioners from recommending the project to the Board of Trade, which confirmed the Order on February 2, 1899.

The favourable petitions from Southall, Hayes and Hillingdon, in addition to that of Uxbridge, had been enough to impel Robinson to make an early start on surveying the road. Local authorities made no excessive demands and the scheme went through with little delay. The Order was made on May 9, 1899. The Southall-Norwood council stipulated only that the LUT should lay wood paving for a specified distance and give the council the right to use the standards for lighting and other purposes. Uxbridge Council made a similar condition for the standards and stipulated that the terminus should be just short of the gates of Osborne and Stevens's timber yard.

The main objection of the county council had centred on the maintenance of the Uxbridge Road, which was only 26 feet wide on average, if the LUT intended to carry heavy goods. Middlesex, moreover, was not disinterested. It now cherished ideas of making itself the tramway and light railway authority for the county and had asked its surveyor to draw up plans of roads suitable for light railways.

The Kew Bridge–Kingston scheme fared differently.

'The proposal to disfigure the locality [Richmond Hill] by a tramway', recorded the *Daily Telegraph*, 'has naturally aroused a great deal of local indignation. . . . However it is only fair to add that a section of the community, chiefly the working classes, are in favour of some plan of the kind.'

The Richmond Borough Tradesmen's Association voted forty-five to one against the scheme. A Richmond councillor declared: 'I believe if the trams come into the borough, the residents will be almost turned out of it themselves by the riff-raff from the East End.'

Sir James Szlumper, future Mayor of Richmond, and an engineer, feared the town would become a second Hammersmith. The idea of the Light Railway Act was to relieve agricultural distress. He could not see how a light railway up Queens Road, Richmond, would do so!

The route had been chosen deliberately to avoid the narrow town centre, otherwise an end-on junction with the Kew Road line would have been proposed. But to meet objections to a route over the hill, LUT supporters proposed a deviation via new streets in the town centre and a widened section of Petersham Road.

The council's objections included the fact that the line would detract from the proposed new Kew Bridge (the future King Edward VII Bridge), and that in 1903 it would be free to acquire the Kew Road horse line itself.

For the time being, Robinson bowed to the Richmond storm and the line was dropped. The Act authorizing the Hounslow extension, Brentford–Hanwell line and power stations was passed on August 12, 1898. By its provisions the LUT was to pay Brentford Council one-third of the cost of widening parts of the long, narrow, congested High Street, to a limit of £5,000, and adopt a special track layout. A Brentford by-pass had first been proposed towards the end of the coaching era, but the High Street, part of the main highway to the west, remained encumbered by drays, carts and animals.

Heston-Isleworth Council gained the right to use lines in the area between midnight and 6 a.m. to carry refuse, road material and night-soil.

Chill winds still blew in Ealing where opposition now centred on possible damage to amenities.

While the LUT Bill for the Acton–Hanwell link was being discussed, Mr (later Sir) Ralph Littler, QC, future chairman of Middlesex County Council, said that Ealing's rateable value in 1864 was £30,000. In the last return it was £290,000. All local authorities concerned favoured the Uxbridge light railway and 3,600 people had signed a memorial in support.

An estate agent said that it was not the trams themselves that depreciated property but the type of property, between £35 and £60 a year, which trams might cause to grow up.

The Opening Shots

Alexander Siemens, director of Siemens Brothers Ltd., said there was nothing dangerous in the Acton–Hanwell route for an electric tramway. Several witnesses even thought that property would appreciate. James Swinburn, vice-president of the Institution of Electrical Engineers, said he did not see why wires should not be put up in Goldhawk Road. The average resident of that road had not a very highly developed aesthetic sense. When he came out of his house in the morning, he did not look at the wires overhead but at the quickest way he could get to the city. Robinson, of Goldhawk Road, must have found that well-meant comment hard to take!

It was argued that frontagers objected mainly to the mass of people trams would bring, and to jerry-building. A Colonel Buchanan said he left Highgate when tramways were laid past his house. Now he would have to leave Ealing. He did not mind seeing workmen in the same train. He did not object to seeing them on trams except when they conducted themselves disagreeably, and on a Sunday. West-enders behaved worse than East-enders.

A valuer stated that people had come to Ealing from Streatham and Wandsworth to escape from trams. Trams would drive away many of the better class from Ealing. His firm was being prevented from letting near the Uxbridge road as buyers objected to the restriction of the minimum value of buildings to £1,600 in view of the possible coming of trams. It was also said that when horse cars first ran on the Kew Road line large houses bringing in £200–£300 a year halved in value.

Sir Montague Nelson, chairman of Ealing council, believed that trams would force out residents. The council's surveyor could not conceive of anything more detestable than the long arms like gallows by which it was proposed to support the wires.

LUT counsel made a shrewd point in asking whether a theatre for Ealing to cost £60,000 was compatible with the rural character about which much had been said.

In addition to a £5,000 payment to Brentford council, the LUT agreed to spend £40,000 on widenings and reconstruction, £20,000 of the sum on widening Half Acre for the Hanwell line. There was about £10,000 to find for street works in Hounslow. Chiswick council demanded a £350 annual wayleave for five years from the passing of the Act, £500 for the next seven years and then £750 until the twenty-one years purchase.

The Duke of Northumberland, the grounds of whose fine mansion Syon House would be visible to passengers riding on the upper deck of

Hounslow cars, was benevolent. (It is recorded that when the route was opened, guests to one of the garden parties at Syon were conveyed in special trams, whose crews wore white shoes and gloves.) A new depot at Isleworth to serve the extension was built on ducal land.

The *Tramway & Railway World* detected ominous signs in the Bill as amended by a Commons committee. The payments to local authorities were most onerous. It was 'more and more the case that whenever a company proposes to extend its lines or to improve its traction appliances its Bill is made the occasion for a universal descent of spoilators. How different it is when a local authority applies for power to work tramways!'

The Act of August 12, 1898, was a partial victory. Only the preamble of the Acton–Hanwell line was not proved. But at least, Hanwell, Southall and Uxbridge would be reached by electric trams, albeit via Brentford.

On the existing system work began in October with track relaying in Chiswick High Road. The old rails were lifted and replaced by new track on a concrete bed. The service was kept going. As a car approached a section being relaid the horses were unyoked. A waiting horse, in the charge of a boy, was harnessed to the car. Keeping a straight course on the near side of the track, it hauled the car to the other end of the section, where it was detached and the pair of horses again attached.

In spite of the Ealing setback, 1898 ended on a hopeful note. Thus the *Chiswick Times* of October 21st: 'The electric car has become a pleasure institution [in the United States] and it has, in the expressive American phrase, "come to stay". . . . We do not see why the electric car should not play some part here in opening up "fresh woods and pastures new" to the workers in our great hives of industry. . . . We should think there is a future for the electric car excursion in Old as there certainly is in New England.'

Another triumph of Robinsonian pertinacity, the Middlesbrough Stockton & Thornaby Electric Tramways opened on July 16, 1898. The *Railway World* said that no better street railway could be found. Major Cardew, after the official Board of Trade inspection, declared he had never seen a more perfect installation as far as the electrical equipment was concerned. Milnes supplied thirty-five cars and had an order for fifteen more. They were double-deck open-top bogie cars with reversed maximum traction trucks and were designed by Robinson. In many ways the system and its cars were the real precursors of the electrified London United.

Its Bill had received the Royal Assent on August 5, 1897. The Imperial

The Opening Shots

Tramways Company had bought the Middlesbrough horse car lines in 1878 and those of Stockton in 1896.

The new system comprised $8\frac{1}{2}$ route miles of 3-foot 6-inch track from Norton Green through Stockton and Thornaby to a terminus in North Ormesby Road. A cross line linked Middlesbrough Exchange with Linthorpe.

The power house had the first Allis cross-compound condensing engines in Britain. BTH contracted for all electrical equipment.

The mayors of Stockton and Middlesbrough praised the admirable way in which the enterprise had been carried out by Robinson and the patience and tact he had shown, often in most difficult circumstances. In this respect also the system was to be a curtain raiser to the LUT.

Shorn now of its Dublin subsidiary, the Imperial Tramways Company was wound up and a new company with the same title formed on September 29, 1898. The capital was £300,000 in £10 shares. Robinson was one of the seven signatories. The new company took over the Middlesbrough–Stockton, the Darlington and Reading horse tramways and shareholdings in the London United and the Corris Railway. It later acquired a holding in the Bristol Tramways & Carriage Co. Ltd.

It should be mentioned that a 600-yard tramway opened on May 13, 1898, deprived Robinson of the honour of inaugurating the first electric tramway in the London area. This was a private line in the grounds of Alexandra Palace which was laid out by a German company, operated by four toastrack cars and in service until September 30, 1899.

Chapter 3

Battle Grounds

Ground for the LUT extensions was broken at Brentford on March 20, 1899. A photograph shows Robinson armed with a formidable roll of drawings and surrounded by a faithful band eager to start work.

Robinson was everywhere. On May 29th he was in Stockton giving a paper on 'Electrical Traction on Tramways' to the Cleveland Institution of Engineers – 'my judgement is that the electric trolley holds the field'. He hoped the LUT would still gain a concession for the Acton–Hanwell line. In deference to the LCC they had undertaken to lay $1\frac{1}{2}$ miles of conduit between Hammersmith and Young's Corner. As the committee of the Royal Society appointed to safeguard Kew Observatory had debarred the company from using rails for earth return they would have to erect double-wire overhead.

Robinson's fresh hope for the Ealing section was founded on the readoption of the section as a light railway, in view of the successful promotion of the Hanwell–Uxbridge line. Ealing Council at once objected that the Light Railway Commissioners were not competent to review decisions of Parliament.

There were stirrings in Surrey, where the Drake & Gorham Electric Power & Traction (Pioneer) Syndicate tried to interest Kingston Corporation in a tramway system based on a capital of £100,000.

Robinson had much bigger fish to fry. Heartened by the success of the Uxbridge scheme he applied for Order for 13 miles 6 furlongs 4 chains of lines, one extending the Hounslow line to Baber Bridge, Hounslow Heath, and another, leaving the Hounslow line at Busch Corner, Isleworth, running to Twickenham, with a branch thence to the Middlesex side of Richmond Bridge, and then forming a loop through Hampton, Hampton Court, Hampton Wick and Teddington.

This was a thrust indeed. At Hampton Wick the loop came within sight of Kingston, and at Hampton and Hampton Court there was the promise of rich tourist traffic in summer.

Battle Grounds

When the scheme was debated by Twickenham Council in December 1898 Robinson said that the LUT would also build a Twickenham–Hounslow cross line when the council wished, provided it succeeded with the present scheme. This was an attempt to oust a proposal by the Drake & Gorham Syndicate for a light railway from Hounslow Barracks to Whitton, Twickenham and Richmond Bridge, and two branches in Twickenham, one to Staines Road at Sixth Cross Road and the other to Wellington Road at Sixth Cross Road.

Both the LUT and its rival proposed to use York Street, a new thoroughfare in Twickenham avoiding the narrow Church Street. The LUT planned to build a new road bridge over the Crane at Cole's Bridge, Drake & Gorham to erect a tram bridge alongside the existing structure.

Twickenham Council was inclined to favour the LUT because of its non-local character.

As regards the Hampton Wick–Hampton Court section, the *Thames Valley Times* of December 14, 1898, welcomed the improvement that the LUT would bring. At present the road surface was of 'a most villainous consistency'. Because of disputed maintenance it was a black shoal of deep, filthy mud.

At Hampton Wick the entrance from Kingston Bridge would be opened out and there would be a 250-foot-long widening up to a depth of 15 feet in the High Street. There would also be many widenings in Teddington.

The projects were inquired into by the Light Railway Commissioners at Twickenham in June 1899. Sam Fay, then traffic superintendent of the South Western, estimated that his company would lose between 1,500,000 and 2,000,000 passengers a year if the LUT lines were built. Waterloo played a cunning game and withheld its formal opposition until the last moment.

The Commissioners, though disposed to favour the LUT then had no option but to find that there would be competition and to direct Robinson to go to Parliament. Higgett of Hampton Council, when Robinson told him of the South Western's last-minute action, exclaimed: 'That is scandalous but quite to be expected of the London & South Western Railway Company.'

Surrey County Council could not make up its mind whether to become a tramway operator, though it got as far as planning about thirty miles of route fanning out from the south side of Hammersmith Bridge.

Middlesex was now ready to project its light railways, estimated to cost £240,000, and was asking local authorities not to proceed with schemes of their own.

The New Electric Traction Company proposed to lay a tramway from Hammersmith Bridge to Barnes, Mortlake and Richmond and build a new street there, but Richmond Council decided that it would itself build any tramways needed in the town.

Rightly trusting to the merits of his co-ordinated scheme, Robinson was undaunted. Armies of men were preparing the existing lines for electric traction. An all-night service of horse cars on both the Kew Bridge and Acton lines was doing well, and it was expected that with electric traction such a service could be run more frequently and cheaply. In July the company announced an interim dividend of 10 per cent, compared with one of 8 per cent in the corresponding half-year of 1898.

The company's Act of August 1, 1899, allowed the overhead system between Uxbridge Road station and 'Askew Arms' and Young's Corner but enjoined conduit between Hammersmith loop and Young's Corner, in accordance with agreement with the LCC and Hammersmith Vestry.

The *Railway World* asked why the LCC should persist in requiring conduit as it had now decided to try out conduit on its own account between Westminster and Tooting. The LUT would have to fit its cars not only with double trolley to satisfy the Kew authorities but also add a plough for use over only 1 mile 3 furlongs 5.9 chains!

The road surfaces were to be supported on yokes 5 feet apart bedded on Portland cement concrete. Excavation was to be 3 feet from the upper road surface. The slot was to be ¾ of an inch wide and manholes with covers were to be provided every 15 feet.

Contracts were placed with British Thomson-Houston for both cars and fixed plant. The *Electrical Review* contrasted the activity of the LUT with the timidity of the LCC. 'While the London United has been actually constructing its West London lines, the Spring Gardens Company has been deliberating as to whether the time had arrived for *experimenting*!'

The LCC had begun in 1892 to take over all tramways in its area – a process which continued until 1909 – with the exception of the LUT Hammersmith lines and some other sections. It began operating them in 1899, with the choice of the type of electric system still undecided.

Ealing Council had resolved, like Richmond, that if it had to have trams, they had better be its own. In November 1899 it gave notice to the Board of Trade for two lines, one from the Acton–Ealing boundary to

Battle Grounds

the Ealing–Hanwell boundary, the other from Uxbridge Road, Ealing, along Longfield Avenue. A councillor said that they had opposed the LUT mainly so that they could keep control of the situation and partly because they wanted to keep the town rural. But Ealing was fast urbanizing. Within ten years it might have 100,000 people and become a second Croydon.

By unremitting propaganda Robinson carefully won over influential opinion in Ealing. In February 1900 the LUT and Ealing Council reached agreement. The price was heavy but the LUT would have to pay more heavily yet before even part of Robinson's ambitions was realized.

The LUT was bound to meet up to £900 the costs of the council's opposition to the LUT's 1898 Bill and also those of the council for its now abandoned application for local tramways! In addition it was to pay the council £500 a year wayleave, plus the usual rates, estimated at £300–£400 a year. Furthermore, it had to widen almost all the main road across Ealing Common and contribute £1,000 towards the cost of moving electric cables. Wayleaves payable to Acton were £400 and to Hanwell, £230.

Roads carrying the tramway were to be repaved with hard wood. The design and construction of the standards were to be subject to approval by the council surveyor and made available for lighting. Ornamental centre poles were insisted on in much of the town.

The LUT had to undertake to complete tracklaying between the north end of the Mall and Christ Church in fourteen days and the section through West Ealing in the shortest possible time. Fares were not to exceed 1d anywhere in Ealing.

These were onerous terms and Robinson was to find Ealing a tough nut yet. And if Ealing could make such exactions, so could other suburbs whose assents were vital. A costly precedent had been established.

With his Ealing flank now secure, Robinson could turn to other fronts. He found new allies in the ratepayers of Teddington. Teddington had had an anti-tram majority on the council, whose members, unlike those of Ealing, retired by thirds annually. When the fight was furious, four returning members were anti-tram men and there was a fifth vacancy caused by death.

Robinson's propaganda machine got busy. Five pro-tram members were elected and Robinson gained his assents in time. At a pro-tram meeting of ratepayers Robinson said that when the LUT came to promote a scheme they found that roads mysteriously narrowed overnight and bridges contracted. Lord Jersey, for the Light Railway Commissioners, had told him that he was sorry that LSWR opposition had compelled them to reject the LUT's light railway application.

A double line of tramway, explained Robinson, was costing £12,000 a mile. Cables and standards added another £14,000–£15,000. The new cars on order would cost £800 apiece, though no horse car had cost more than £220. They were spending £100,000 on the new power house and depot at Chiswick. The service would operate 16 hours a day, at 5-minute intervals for 12 hours.

The hon. secretary of the Teddington ratepayers' committee had a forcible announcement printed in the local press in February 1900. Entitled 'Why Should Teddington Wait?' it reiterated the agreement of 1899 by which the LUT undertook to widen Teddington railway bridge, carry out other road improvements, and wood-pave the track – in some places the whole road – at a cost of £30,000. It showed how, when the LUT was forced to go to Parliament with the scheme, members of the council tried to introduce a clause to give them authority to purchase at 'an old iron price' instead of at the fair market price. The author could not understand how they could think of losing £30,000 worth of improvements, at no charge on the rates, for the offchance of gaining for the ratepayers of 1925 'shadowy rights of purchase' of lines they might not wish to buy.

The trams were described in the announcement as being so elegant and up to date that in many instances they did away with the cost of hiring private vehicles. 'It has been proved conclusively that the introduction of a modern electric tramway service improves the value of property all along the route, and that it promotes and stimulates the building of residential quarters in its immediate proximity.'

In south Middlesex the LUT already possessed the London ends of the Bath and Oxford roads and with these as bases was extending. Middlesex tried to induce the LUT to dispose of its undertaking and sought to baulk its new schemes, but could not persuade all local authorities to form an anti-LUT league, doubtless because the LUT promised to be a fat milch cow.

Robinson struck back and gave notice of his intention to invade north Middlesex, whereupon the council gave in and agreed on a 'sphere-of-interests' demarcation line. The arrangement, ratified on November 28, 1901, confirmed the LUT in possession of Middlesex south of an imaginary line drawn north of Uxbridge Road.

In December 1901 Acton council passed a resolution deploring the

Top: Type X No. 117 on Hanwell service at Brentford Half Acre. *Centre:* Type W No. 184 follows a haycart through narrow, congested Brentford High Street. *Bottom:* Brentford Canal Bridge, with Nos. 4 (type Z) and 182 (type W).

Top: Cars 7 and 19 of type Z pass outside Hounslow depot. *Centre:* At Hounslow ('Bell') terminus, soon after the opening from Kew Bridge in 1901. Note addition of 'Hounslow' to original route lettering on upper deck panel. (*Courtesy Hounslow Library.*) *Bottom:* Hounslow Heath ('Hussar') terminus. No. 61 of type Z (later Y). (*Courtesy Hounslow Library*)

A busy traffic scene at Shepherds Bush in May 1903. (*London Transport*)

Holiday crowds at Shepherds Bush in the early days. (*London Transport*)

Top: Type X No. 143 on Kew Bridge–Shepherds Bush Service rounds the corner in Goldhawk Road, by Stamford Brook Road. (*Courtesy Hammersmith Library.*) *Centre:* Type X No. 125, seen here eastbound in Acton, was sold to Blackpool in 1919. (*Courtesy R. B. Parr.*) *Bottom:* Type T No. 335, on left, passes type X No. 148 in Acton High Street. Note interlaced track in foreground. No. 148 ultimately became Hanwell depot works car. (*Courtesy Acton Library*)

Battle Grounds

action of Middlesex County Council in agreeing to surrender its tramway 'rights' over much of west Middlesex without consulting the local authorities. But those authorities were ready to go along with the LUT at a very stiff price.

The viewpoint of Middlesex County Council was made plain in a subsequent (January 1902) memorandum issued by its chairman, Ralph Littler. Ironically, as we have seen, Littler had appeared for the LUT in 1898. He contended that Parliamentary debates on the Light Railway Act showed that it was intended to apply solely to the building of light railways, not to what were urban tramways in all respects save legal liabilities and immunities.

Littler said that the LUT, finding no preamble in the Act, nor any definition of light railways in it, determined to make use of it to gain powers for what should have been put forward as a tramway. It was the success of this manoeuvre that prompted Middlesex Council to resolve on February 23, 1899, to make use also of the provisions of the Act and instruct the county engineer to prepare a scheme.

The London & South Western Railway was consistent. It told Middlesex that it would oppose council light railways in Heston, Isleworth, Teddington and Hampton in the same way as it had opposed the LUT's Thames Valley light railway scheme.

Although, as we shall see, the LUT Thames Valley scheme passed in the form of a tramway, and local authorities did not co-operate with the county council in objecting to LUT plans, Middlesex considered it desirable to gain possession of some roads in the south-west of the county which the LUT did not already occupy or was not yet authorized to occupy.

For this reason Middlesex sought powers for a network of light railways in May 1901, to serve an area bounded by Staines, Cranford, Isleworth, Twickenham, Hampton and Laleham. Included were: lines from Staines to join the authorized LUT line at Hampton; Staines–Baber Bridge and then across fields to Hounslow Barracks (now West) station; Hounslow– county boundary near Colnbrook; Southall–Hounslow, with a branch to a proposed generating station at Bull's Bridge; Twickenham–Staines Road (Kingston Road/Vicarage Road junction); and a short line at Hounslow. Powers were also sought to make junctions with the Metropolitan District Railway at Hounslow Barracks and with the projected Staines & Egham Light Railway.

As the LUT's proposed light railways in the same areas duplicated some of the council's, Littler 'suspected that their plans were copied from ours'. This was surely an unjust inference seeing that there was only a limited number of roads which were, or could be made, suitable for light railways.

It was the continuing attitude of local councils and unrelenting pressure from the LUT which, alleged Littler, made him decide to strike a bargain with Robinson. (This was the sphere-of-influence agreement ratified in November 1901.)

Middlesex had already decided, in January 1900, to approve of the principle of building light railways which it might have power to lay down and then lease to a company to equip and work. This was not the sort of arrangement likely to appeal to Robinson but it would soon be applied with success outside the agreed LUT sphere of influence under the term of a lease agreement with the Metropolitan Tramways & Omnibus Co. Ltd.

The LUT won a notable victory at Hammersmith, where the council had protested on finding that its roads would have to be broken up to lay the conduit on which the LCC insisted. The LCC at last gave way and allowed overhead instead.

The construction of the Hanwell–Uxbridge light railway in no way differed from that of the tramways. The Hanwell–Southall section was put in hand at the same time as the building of the Acton–Hanwell tramway. Although there was still open country between the River Brent and Southall, double track was laid in justifiable expectation of heavy traffic. Unfortunately, similar foresight was not shown with the rest of the extension, which was laid mostly as a single line with loops, a circumstance which prevented its full development in later years.

Centre poles were used in the centre of Southall as in Ealing. Elsewhere, as on other new lines, the side standards on the new lines were steel tubes of three types, bedded 6 feet in concrete and slightly raked. The base rested on a cast-iron plate. They were required to withstand strains of 700, 1,000 and 1,500 pounds respectively applied 2 feet 6 inches from the top, with a temporary deflection of not more than 6 inches.

All structural, outdoor, architectural and technical work was designed and carried out by LUT staff. The engines, dynamos and other technical equipment were supplied by contractors to LUT specifications.

The trolley wire was hard-drawn copper wire weighing 960 pounds/ 1,000 yards and suspended not less than 21 feet from the ground except under bridges. Routes were divided into half-mile sections, with section

Battle Grounds

boxes between each. Feeder cables were lead-covered and paper-insulated, in wrought iron, cement-lined ducts laid under footpaths or between the tracks. Rails were of 92½ pounds a yard section in 36-foot lengths and double-bonded.

When a House of Commons committee presided over by Sir U. Kay-Shuttleworth began to examine the Thames Valley lines Bill, Balfour Browne, QC, put the LUT case with skill. He said there was no opposition to the Hounslow Heath extension. Elsewhere, the only opposition came from Middlesex County Council and certain frontagers affected by the proposed Twickenham–Richmond Bridge line. Local authorities had been won over by the promise of street improvements to the tune of £118,000.

Evidence was given that Twickenham's bus and train services were inadequate, though Richmond Road objectors retorted that they were amply catered for by buses. They also pointed to the crowded state of the road and the added danger from the construction of a tramway along it, with a terminus near the foot of Richmond Bridge. But the owner of many houses on the route thought that trams would benefit his properties, of which nine had an annual value of £55, seven of £65 and four of £75.

Balfour Browne hoped that the Bill would pass with the Richmond Bridge line, which, though not an integral part of the scheme, would be most important in the future.

To vague suggestions by Middlesex Council that it would some day build tramways in conjunction with the LCC, Robinson replied by stating an existing traffic of 10,000,000 passengers a year and to a probable 40–50,000,000 if the proposed lines were completed.

The committee decided that Middlesex had no *locus standi* as regards the desirability of the construction of lines and could oppose only in the mater of interference with roads and bridges under its control. It was prepared to sanction the Richmond Bridge line if the LUT would improve curves and provide a 9-foot 6-inch width between the outer rail and footpath where possible.

The gaining of the Act on August 6, 1900, was a great triumph for Robinson. It authorized the Acton–Hanwell, Hounslow–Baber Bridge and Busch Corner–Hampton Court loop lines, as well as the Twickenham–Richmond Bridge branch, a short link line in Boston Road, Hanwell, and doubling between Acton Vale and Acton High Street. The Thames Valley line would bite deep into LSWR preserves, tapping an area rich in potential tourist traffic and affording a springboard for further promising incursions.

The first of the LUT's electric trams was shown at the International Tramways & Light Railways Exhibition which opened at the Royal Agricultural Hall on June 22, 1900. It was one of 100 bogie cars being supplied by Hurst Nelson, with BTH electrical equipment. The general design owed much to Robinson himself and was a development of the bogie type already proven on Tees-side, itself based on the 4-wheel Dublin Southern and Bristol designs.

It came as a revelation to Londoners, used to small horse trams and buses. It was lavishly equipped, with longitudinal seating for thirty-nine inside and transverse seats for thirty outside on the open top, with its ornate grillwork. The saloon seats were spring-cushioned and covered with plush moquette, and the ceiling was panelled in birdseye maple. There were electric push buttons as well as electric lighting. The car was mounted on Peckham maximum-traction bogies.

The first drawings of LUT electric cars show a vehicle with twin trolleys, mounted one on each side of the upper deck, for use with the double trolley wire which Robinson originally thought was inescapable.

Robinson had wanted to use the City of London arms on the car sides but the City Corporation withheld sanction as it had no connection with the LUT. He then settled for what Mr Charles E. Lee has termed a 'mutilated version'. The sword of St Paul and the words *'Domine dirige nos'* – hardly appropriate to a tramway company – were deleted and the badge was a shield bearing the cross of St George with griffins as supporters.

Financial results for 1899 confirmed the wisdom of an expansionist policy. Gross receipts rose by £6,824 to £54,369, making a net profit of £13,271. A final dividend of 6 per cent on the preferred and 10 per cent on the ordinary shares was declared. The Imperial Tramways Company increased its holding in the LUT by 15,000 6-per-cent preference shares.

The battle of Kingston was about to be joined. Tramway promotion was no novelty to Kingston and the first proposals were launched in 1871 by the Kew, Richmond and Kingston-on-Thames Tramways for a line from Richmond to Kingston Market Place. The Bill was thrown out. The next attempt, also rejected, came in 1883 when the Kingston Corporation and the Surbiton Commissioners jointly promoted quite a comprehensive system. It comprised tramways from Thames Ditton along Portsmouth Road to Kingston Market Place, thence by a loop via

Battle Grounds

Fife Road, Kingston Station and Clarence Street; along Surbiton and Claremont roads to Surbiton station and Ewell Road; a link from Surbiton station along Victoria and Brighton roads to the Portsmouth Road line, a line from Clarence Street along London Road to the Liverpool Arms, Norbiton; and another line along Eden Street.

In January 1899 the General Purposes Committee of Kingston Corporation threw out proposals by Drake & Gorham for tramways in the area but decided to call a meeting of all local authorities affected by the scheme. Councillor T. Lyne gained support for his view that any tramways ought to be built by the Corporation. Both BET and Greenwood & Batley contended to build light railways. BET bought the Greenwood & Batley rights but to no avail, the Light Railway Commissioners rejecting the scheme, largely because of Surrey County Council opposition.

By June 1899 the Corporation had prepared its own plan. It was a sound enough project, save that most of the lines, except in Clarence and Eden Streets, were to be single with loops. Routes were proposed from the Malden boundary to Norbiton Church and along London Road and Clarence Street to join the authorized LUT line in Hampton Wick; from Norbiton up Kingston Hill as far as the 'George & Dragon'; from Eden Street junction along Richmond Road to the Ham boundary and along Eden and High Streets and Portsmouth Road to the Surbiton boundary; from Eden Street via St James's, Penrhyn, Surbiton and Claremont roads to Surbiton station, with a link between Portsmouth Road and the Surbiton Station route via Surbiton Road.

Members of the Corporation went to Dover in July and were most impressed by the electric tramways opened there in 1897. They were heartened by a proposal of the London County Council for a light railway from Clapham via Wandsworth and Putney Vale to the county boundary at the Beverley Brook, whence there was only a $1\frac{1}{4}$-mile gap to the Corporation's projected Kingston Hill terminus. In good hope therefore the Corporation put forward its Parliamentary Bill on November 20, 1899.

On March 27, 1900, after two days' consideration, Parliament rejected the Corporation Bill because it was not comprehensive enough. It is hard to see how a purely municipal scheme could have been more embracing. The corporation and civic interests were disappointed and there was strong feeling that Kingston should try again.

At the right moment Robinson came forward with a wide-ranging scheme which not only duplicated almost all the Corporation's plans but also extended to Tolworth, Thames Ditton and Molesey and threw in a branch to Richmond Park Gates.

The full list of routes in this part of Surrey for which authority was sought was:

Hammersmith – Putney – Putney Heath – Kingston Hill – Kingston – Hampton Wick (to join authorized line)

Kingston (Eden Street, St James's Road, Penrhyn Road) – Surbiton (Claremont Road, Victoria Road, Brighton Road, Portsmouth Road) – Long Ditton – Esher (Broom Hill)

Kingston (Norbiton Church) – New Malden Fountain, via Cambridge and Kingston roads

Kingston (Eden Street – Richmond Road – Kings Road – Richmond Park Gates)

Surbiton Station – Tolworth, Red Lion, via Ewell Road, with branch via Ditton Road and Brighton Road to Hook (Epsom Road)

Long Ditton – Thames Ditton – East Molesey – Hampton Court (to join authorized line).

With this ambitious project another great wedge would be driven into the LSWR's suburban business. Kingston, which in the 1830s had spurned a main line railway, would at least be placed on a tramway trunk route. More rich areas of residence and resort would be tapped.

The well-timed move sharpened controversy in Kingston. One group urged negotiation with Robinson so that the Corporation might secure the ownership, if not the operation, of lines within the borough. Extremists pressed the Corporation to oppose the LUT and go to Westminster with a fresh scheme.

J. Edward Waller of Kincaid, Waller & Manville had reported favourably to Kingston Corporation on its own scheme. 'Kingston is the key to the whole position. . . . If you retain the key you will be able to make much better terms than you could possibly expect were you now to enter into negotiations with the one company who for the moment is in the field.'

Surbiton Commissioners decided to go along with the LUT. Maldens & Coombe agreed not to oppose the LUT after Robinson had given in on certain points. One Malden councilman considered that tramways endangered cyclists – the *Tramway & Railway World* said that cyclists benefited from tramways if the roads were wood paved, as between

Battle Grounds

Hammersmith and Hounslow. Esher opposed. Molesey was ready to make terms.

For the present outside the Robinson orbit, though sure soon to be attracted, Wimbledon decided to promote a Bill for a tramway from Wimbledon station to the LCC boundary at Tooting.

The *Surrey Comet* ran editorials advocating local tramways or a joint tramway committee of north Surrey authorities. In its July 22, 1900, issue it said ' . . . the tramway company's Bill has been engineered with wonderful adroitness. The company's agents are skilled to a remarkable degree in the arts of tramway promotion, and they succeed in getting round the local authorities and the majority of the inhabitants by the energetic use of methods to which, we must admit, we were previously strangers'.

In November 1900 the tramways committee of Kingston Corporation issued a 14-page statement stressing that it was desirable for the burgesses to retain control of tramways in the borough. The corporation strongly favoured having tramways in the town but emphasized that it had to have parliamentary powers even if it leased the lines. It also disclosed that in March 1899 Robinson had said that he would negotiate for a lease of lines in Kingston once the corporation had gained its powers. Now that the LUT had put forward its own scheme the ratepayers ought to share in any profits the LUT expected to reap.

A doggerel poem which went the local rounds accused the LUT of caring only for feathering its own nest. It began:

> 'Who wants the tramway here?
> 'I,' said Cock Robinson, 'What, ho!
> 'My tramway shares must upward go,
> 'I want the tramway here.'

Ratepayers met to consider resolutions authorizing the corporation to reapply for tramway powers and oppose the LUT. The meeting hall overflowed and the proceedings were a series of disorderly interruptions.

Councillor Lyne, by this time Chairman of the Tramway Committee, had to sit down half-way through his speech. There was a scandalous incident when a Dr Finny ('For he's a jolly good fellow'), supporting the motion, produced a telegram which he alleged he had received. It ran: 'Refrain from meeting and 500 preference debentures [*sic*] are dependable.' Wild cheers changed to hisses when the telegram was seen to be unsigned and a solicitor called Wilkinson denounced it as a forgery.

Robinson retorted at once with an answer which showed his concern that no mud should stick to the LUT.

Although the resolution was lost by a large majority, another authorizing the corporation to oppose the LUT Bill was passed.

Robinson could now afford to ride the Kingston storm. The passage of the Thames Valley lines Bill had prompted an editorial 'Tramways Triumphant' in the July 21st issue of the *Middlesex County Times* as follows: 'The little knot of nor'easters who rule the roost at the Westminster Guildhall had stuck at nothing to secure the rejection of the Bill. They have strained every nerve and spent county money like water in an effort to wreck the company's scheme, or, alternatively, to rob the local authorities of the benefits they had secured for their districts by agreement scheduled in the Bill.'

At a Middlesex County Council meeting in November, Alderman James Bigwood spoke of districts that had listened to the voice of the charmer. 'And quite right too', came a retort.

Chapter 4

Electrics at Last

A Richmond resident wrote to tell Robinson in November 1900 that he had given up his first-class season ticket between Richmond and Waterloo as he thought £16 a year exorbitant. To discourage traffic via the Central London Railway, he said, the South Western had taken off several trains between Richmond and Shepherds Bush. Although it was rather quicker by train than by horse tram between Gunnersbury and Shepherds Bush, the trains could not be relied on as they were sometimes 15 minutes late.

Robinson's reply included the information that the fare by electric tram would be 2d and the journey would take about 20 minutes. They hoped to get powers to cross Kew Bridge, when it was rebuilt, go through Richmond and over Richmond Bridge.

The year 1901 opened with Kingston still fighting, but losing. Only 1,123 ratepayers backed the corporation's tramway scheme compared with 3,701 for the LUT. A postcard poll in Malden and Coombe also favoured the LUT. Malden Council agreed to oppose any competing scheme but expected the LUT to meet its costs in so doing.

Finally Kingston Corporation resolved to complain to Parliament that the LUT was not complying with Standing Orders. It was a last card to play if the LUT did not offer better terms.

Work was going ahead rapidly on the electrification of the existing lines and the erection of the large works and depot at Chiswick alongside the old West Metropolitan depot and adjacent spare land.

The Kew Observatory impasse remained. Robinson told a *Daily Chronicle* reporter that the Kew instruments could be replaced elsewhere for £100 and supervised for £1 a week. Instruments in Paris had been removed out of the range of possible electrical disturbance.

In the Kew affair the Board of Trade was in the hands of the Treasury and the Royal Society. The LUT declared it would rather pay interest on capital than agree to the Kew authorities' prohibitive terms.

London United Tramways

In January members of Chiswick Council visited the power station and depot. Robinson pointed out what would be a new London landmark, the power station's steel chimney rising 260 feet from the foundations and said to be the tallest steel chimney in the world. American workmen had put its steel in place in thirty days, a record. British bricklayers finished the foundations and ground work in a fortnight. 'They can do it when they like,' commented Robinson.

The foundations of the chimney went down 28 feet to the blue clay. The base was of red brick and Portland stone and the steel plating of the stack was fixed to a cast-iron baseplate.

The building, 106 feet deep and with a frontage of 154 feet to the courtyard, was divided into a boiler room and an engine room. The boiler room was mainly a steel frame structure to carry the bunkers, which could hold 500 tons of coal, the flue, the 360-tube 'economizer' and the feed water tanks. Each of ten water-tube boilers, fired by mechanical stokers, could evaporate 11,000 pounds of water an hour at 150 pound psi. Three 30-foot bores yielded 10,800 gallons of water an hour.

The engine room was 60 feet wide. It was served by an overhead travelling crane and housed two vertical cross-compound condensing engines with cylinders 22 inches and 44 inches by 42 inch stroke. Each engine was connected to two 250 kW. generators supplying power to the overhead. In addition there was one 500 kW. generator, to be increased by two 1,000 kW. three-phase sets, to supply power to substations. The switchboard contained 27 marble panels bolted to angle-iron frames.

The horse car depot at Chiswick, set well back from the High Road, had consisted of a 3-track shed accommodating twenty trams, stables, loose boxes and other buildings, with a siding that had been used for the trials of the Lineff system.

The fine new car shed had 5 tracks, each 390 feet in length, and a 6-track repair shed with 1,500 feet of track. All tracks had inspection pits. The ample accommodation also included a machine shop and a traverser.

The capacity compared with 7 tracks at Acton depot and 10 at Hanwell and at Hounslow.

Robinson said it was difficult to find room for the horses needed until the electric cars began running. Between 300 and 400 electric cars would be needed as the service demanded 6 per mile. They were going to have 96 miles of line and when they got to Hampton they might some day want something like 1,000 cars.

Electrics at Last

The designer of the power station was William Curtis Green. The exterior and much of the interior were finished in ornate style. Outer walls were of brick and stone. The design of the stone plaque over the main door included female figures representing Electricity and Locomotion, the one with her feet resting on a globe, the other with them placed on a relief of one of the tramcars. The initials LUET formed the centre of the group between the figures.

Inside was an elaborate curved staircase leading to the instrument gallery. Both staircase and gallery incorporated intricate designs in gilded ironwork.

This magnificence was mostly hidden from view by three houses in the main road. A rather narrow entrance, traversed by a single track, was flanked on the west side by the headquarters offices. They included rooms for Robinson and his staff, the board room, the traffic superintendent's office, offices for the inspecting staff, and reading rooms and other accommodation for the inspecting staff.

The board at this time comprised, in addition to Robinson, George White (Chairman), Edward Everard, Hugh Charles Godfray and Samuel White. George White was also Chairman of the Bristol Tramways & Carriage Co. Ltd and since 1897, of the Imperial Tramways Co. Ltd. Everard was a Bristol printer and publisher, Godfray was also a director of the Imperial company and Samuel White was managing director of the Bristol company.

Robinson had determined to secure the great prize of Richmond for his electric empire, though the council was proving difficult, if not adamant. In February the LUT produced a lavish 28-page booklet *Electric Tramways for Richmond*. It proclaimed that the LUT would eventually cover nearly 100 miles of route from Regents Park and Kensington to the Staines and Bath roads, to Kingston and Surbiton, and to Putney, Barnes – and Richmond. Cars would run for 18 out of the 24 hours at statutory maximum fares of $\frac{1}{2}$d a mile. 'What can they know of the possibilities of the Electric Tram in London who know only the London Horse Tram of today?' asked the publication.

In its issue of February 16 the *Richmond & Twickenham Times* was at least able to commend Richmond Corporation for deciding not to appoint a tramway committee. 'Any attempt on the part of small suburban authorities to substitute little local systems for large general ones, will certainly be time and labour and money wasted', it wisely warned, adding that Kingston had made itself a laughing stock in this way.

Such a committee was nevertheless soon formed. Almost at once it debated a system to be promoted by Richmond and Barnes councils and the Richmond Electric Lighting & Power Company. It decided against the scheme and also to continue opposition to the LUT pending the result of negotiations.

Richmond and the LUT reached agreement in April. The LUT would adopt conduit in Richmond, agree to retain the purchase clause, pay an annual wayleave of £500, charge a maximum fare of 1d within the borough and 2d to Putney or Hammersmith, pave and maintain the tramway area with wood blocks, and widen streets at an estimated £30,000.

In March the Board of Trade inspected the LUT's electrical equipment. Horse cars were halted for up to two hours between Hammersmith and Kew Bridge to allow thirty electric trams to be tested. R. T. Glazebrook of Kew Observatory watched the measuring instruments at the power house and noted the power fed into the line. The data were compared with magnetometer readings at Kew to ascertain any effects.

A bold stroke by Robinson at last settled the Kew deadlock. The opposing parties were brought together before the Chancellor of the Exchequer and it was agreed to move the observatory to the Eskdalemuir in Dumfriesshire, with the LUT meeting most of the cost.

On March 30th, Lieut-Colonel H. A. Yorke of the Board of Trade inspected all the electrified lines, except the Hounslow extension. The surveyors of Hammersmith, Acton, Chiswick and Brentford declared themselves highly satisfied and urged an early opening. In this general euphoria Robinson cannot have been greatly dismayed in May by being fined 10s for allowing unlicensed drivers to work – Scotland Yard had delayed an application for licences for 200 men, who had been trained by motormen brought in from Stockton and elsewhere.

With Board of Trade sanction given on Wednesday April 3rd Chiswick depot was ready to start public service by electric cars at 7 a.m. next morning, in time to profit by the Easter holiday.

The *Chiswick Times* of January 25th had praised the cars. '. . . the woodwork of the roof and sides . . . reminds one of the fittings in the saloon of a handsome yacht. Stairs are straight instead of rounded, so ladies will appreciate the change. By touching buttons, like those on electric bells, the ventilators can be opened or shut'.

Local reporters approved of the elegant form and finish of the standards which had lessened opposition to the overhead system. But three

Electrics at Last

was an ominous note too. 'The probabilities, all the same, are that the trolley is destined, at some period or other, to give way to a more sightly system. Street locomotion is in a transitional stage, and it is tolerably safe to predict that in 20 or 30 years' time the novel methods of today will seem as antiquated, and be nearly as obsolete, as the horse traction of yesterday and today.'

The new system met the demands of Easter triumphantly. All eighty-five cars so far delivered were in service, at 500 yards intervals. Only two developed hot boxes. They carried 385,000 passengers over the Easter holiday, a superb achievement. Construction gangs had to be brought in from the extensions to control the crowds. Even these figures were surpassed over the three days of Whitsun when 443,000 people were carried, with a maximum of more than 2,100 per car on the Monday!

The speed of the new cars discomfited other road users and there were soon wordy exchanges between motormen and other drivers who refused to give way. The unusual appearance of the large, ornate tramcars amazed the unsophisticated. An Irish labourer is said to have observed: 'Well, bedad, I've seen tramcars pulled by 'osses and driven by steam but this is the first toime I iver saw one pulled by a fishing rod!'

The gongs caused complaints. So did the noise of running, a charge which would be constantly levelled through the years. One Chiswickian wrote: 'Why run at speed so late that it is impossible to sleep till after one a.m.? Why not one or two gongs instead of continuous gonging?' Another deplored the 'awful noise', adding that on the preceding night, about 11 p.m., six cars all going in the same direction passed him in a distance of 300 yards, each with an average load of five passengers. Yet another resident found the uniform colours of the cars confusing! This charge at least would not be valid long.

Even the smart conductors came in for criticism. 'I was carried past a stopping place, and on protesting I was coolly informed by the Germanized-looking uniformed conductor that I could take a penny bus back! I shall take his advice and stick to the safe, clean, comfortable and convenient omnibuses which allow people time to alight where they want to.'

Teddington had already lodged a formal objection to the LUT's Thames Valley Bill but in May Twickenham decided to withdraw opposition if given the option of purchase independently of other authorities and if the LUT promised not to lay cables without a tramway.

Chapter 5

Junkets – And Hard Bargains

In *The Times* of April 29, 1901, Robinson expressed his interest in the proposal to amend the irksome Standing Order 29 of the Tramways Act of 1870. This was the order that provided that a tramway Bill promoter had to show formal assents of the local bodies concerned, representing not less than two-thirds of the mileage of continuous lines proposed. If, for instance, a line were proposed to run through the area of authorities A, B, C, D and E, the scheme could be wrecked by D's opposition, even if the rest had assented.

Robinson wrote that it had become almost impossible to get assent anywhere without accepting not only prohibitive but demoralizing terms. If a company yielded to local demands could it be accused by interested parties of having bribed the authorities. Tramways ought to be on the same footing as railways, gas, water and other like undertakings. 'No established company would be so foolhardy as to undertake the very expensive business of promoting a tramways extension Bill unless there were some public demand.'

In another letter to the same journal, on May 17th, Robinson commented favourably on a letter from E. J. Halsey, chairman of Surrey County Council, who had written of the 'blackmail' being levied from companies promoting electric tramways. Robinson contrasted this with a letter from the chairman of Molesey Council displaying ignorance of the Tramways Act and Standing Orders. There was no demand in them, said Robinson, that there had always to be 9 feet 6 inches between the outside rail and the kerb, a width Molesey had laid down as one of the conditions on which it would agree to tramways.

The fine weather at the end of April and beginning of May brought heavy traffic to the electrified routes. Kew Bridge terminus on Saturday and Sunday afternoons saw a continuous stream of laden cars arriving and departing.

Wednesday, July 10, 1901, was a great day in the annals of West

Junkets – And Hard Bargains

London. The LUT had chosen it for the formal inauguration of both the converted and newly-built lines, possibly out of compliment to Ealing, incorporated on the same day.

Robinson and his fellow officers spared no trouble or expense in making the occasion truly worthy of the completion of London's first public electric tramway system. All along the new extension from Acton across Ealing Common to Ealing, on to Hanwell and then through almost open country to Southall, a line which had been built at the rate of almost a mile a week, the standards were gay with flags and shields.

For the distinguished guests there were nine of a new batch of cars, finished in white and gold and adorned with pink roses and smiles. And what a guest list! The Marquess of Lansdowne, the Earl of Onslow, Earl Grey, the Earl of Rosse, Lords Herries, Rothschild, Revelstoke, Hillingdon, Mountstephen and Farquhar, the Rt Hon. Arthur (later Earl) Balfour (leader of the House of Commons), Sir Hiram Maxim, Sir Edward Reed, Sir William Preece, Sir Ernest Cassel, Sir Benjamin Baker were among the titled. Then there were the American magnates Charles Tyson Yerkes and Pierpont Morgan, Jr (both soon to be more closely associated with the LUT), Granville Cunningham, General Manager of the Central London Railway, and W. J. (later Sir William) Bull, MP for Hammersmith and a friend and champion of Robinson.

Car No. 105 led the caravan. The start was slightly marred by a furniture van breaking down on the track. But a band playing on Shepherds Bush Green enlived the minutes of delay. After that, all went well on the inaugural run to Southall. On the return trip the cars stopped at Ealing Town Hall, where Lord Rothschild declared the lines open and congratulated Ealing on having a scientific and comfortable system of locomotion brought to its doors.

The cars reversed at Shepherds Bush and then went along Goldhawk Road to Chiswick, where, after inspecting the power house, the party lunched in the depot which the magic of Messrs Ritz and Echenard of the Carlton Hotel had transformed into a 'bower of roses garlanded with smilax and draped with old gold cloth'. The Red Hungarian Band of the Carlton Hotel played fifteen pieces, including Berger's Electric March.

During the sumptuous luncheon, for which the Carlton was also responsible, Balfour proposed 'Success to the London United Tramways'. 'I look to this vast enterprise', he said, 'not merely to diminish the evils of metropolitan congestion but greatly to add to the highest pleasures of life of the metropolitan inhabitants.' He had been told that when the system was complete it would be carrying 150,000,000 passengers. 'Enterprises like this', declared Balfour, 'do far more to solve the housing problem and ameliorate social conditions than any legislation.'

George White said that in a year the number of passengers carried by the horse cars was less than that carried by the electric cars in three months. In 1900 the nine miles of horse tramways had carried 8,000,000. The company lost £3,000 a year by excluding advertisements from the cars but they believed the policy was wise. [That view would soon change]. White declared: 'All this great work which has been accomplished would not have been carried to such a successful conclusion without . . . the existence of a master mind.'

Robinson told his hearers that what they had seen was a prelude to greater things. Even though they might not realize Mr Balfour's plan for radial roads to relieve congestion, their own roads were radiating east, west, north-east and north-west. In due time, he hoped, they would go south and south-west too. He permitted himself a little joke at White's expense. 'Why was George White? Because, of course, Robinson Crusoe!'

Robinson had arranged for a film to be made of the inauguration and shown that evening at the Empire Theatre, where he reserved all the stalls so that as many LUT staff as possible could see it.

The inauguration spurred the LUT publicity department to produce a handsome illustrated book containing historical details of the districts served and excellent descriptions of the development of the company and its equipment.

The public service on the Southall route which immediately followed was marred by some more or less minor accidents during the following days. A tram collided with a bus in Shepherds Bush and some passengers on the upper deck were injured. Only a tram standard kept the bus upright. Another tram ran into a laden coal cart in Acton, yet another into a brougham on Ealing Common. Some cyclists came to grief through trying to turn too sharply as they crossed the metals. Such mishaps were caused largely because other road users were unfamiliar with the nature of the new railbound beast that had come to share the way with them.

On the night of June 24, 1901, Clifton Robinson junior had driven a trial electric car to Hounslow. Horse cars had run over the extension as early as August 9th and 10th, 1900, to test the gauge, and the line had been authorized for horse traction on August 11th.

Top: Type X No. 132 on interlaced track at west end of Acton. (*Courtesy Acton Library.*) *Centre:* No. 112 splashes across Ealing Common in a deluge on 15 June 1903. (*Courtesy G. L. Gundry.*) *Bottom:* Type X No. 138 on Southall service in Ealing Broadway. (*Courtesy Ealing Library*)

Top: Ealing Broadway, looking east. Type Z (later Y) No. 78. (*Courtesy Ealing Library.*) *Centre:* No. 210 of type W (later U) passes along Ealing Broadway on its way to Uxbridge. (*Courtesy Ealing Library.*) *Bottom:* No. 223 of type W crosses the Brent at Hanwell en route to Shepherds Bush.

Top: Type T No. 336 on Southall–Shepherds Bush service at Southall Town Hall. (*Courtesy Leonard and Marcus Taylor.*) *Centre:* Rural charms at Hillingdon on the newly-opened Uxbridge route. (*Courtesy G. L. Gundry.*) *Bottom:* Early days at Uxbridge. Type Z (later Y) No. 15 at the terminus.

Top: Type W No. 212 at Richmond Bridge terminus. Local opposition to the use of narrow side roads to the left of the picture caused the withdrawal of the LUT's first trolleybus proposal. (*Courtesy The Reverend P. W. Boulding.*) *Centre:* Cars in King Street, Twickenham, looking east. *Bottom:* Stanley Road Junction. Type W (later U) No. 295 on left diverges for Teddington and Hampton Court; type W. No. 258 heads for Hampton Court via Hampton.

Junkets – And Hard Bargains

Public service with electric traction began between Kew Bridge and Hounslow on Saturday July 6th. The first workmen's car ran at 4 a.m. and the first ordinary car at 7.30. The headway was 5 minutes and the last car ran at 2 a.m.

Although Isleworth and Hounslow people may have seen electric trams at Kew Bridge, the novelty of seeing them at Hounslow impelled attention. The extension was an instant success. One City man said that he now went to the Bank by LUT and Central London Railway well inside 1 hour 25 minutes and had comfort all the way.

The hundreds of employees of Pears at Isleworth forsook train for tram. 'It's the heaviest blow the South Western Railway have had,' exclaimed one tram passenger as an almost empty train crossed the bridge at Spring Grove. 'About time something woke it up', said another. 'We need not mind now if we miss their one-an-hour Gunnersbury trains, for we can get through every few minutes.'

The first Sunday of operation brought thousands into Hounslow. The bottleneck of Brentford was a serious nuisance, however. The unusual provision of a succession of crossovers to allow wrong-line working – there were five facing and five trailing in three-quarters of a mile between the Royal Brewery and Brentford canal bridge – was a palliative but to the end Brentford High Street remained one of the LUT's biggest incubuses. Robinson said later that this arrangement worked well in the first year. Then, interested parties saw a chance of getting the whole street widened at LUT expense.

Even so, Brentford in general welcomed the trams, though the noise of the trolleys disturbed some dwellers in the High Street as the wires ran close to their bedroom windows!

In February 1901 the company had bought from the Duke of Northumberland a plot of land between Isleworth and Hounslow called Three Rood Piece. A 10-track depot with 1,250 feet of track was built, with a substation alongside. A permanent way depot and stores was built at Brentford.

Robinson well understood the power of the press and usually got a good one. The papers did the inauguration proud. One London reporter at least was lyrical. 'The racegoer of 1904 will be able to run down by tram to Epsom Downs on Derby Day at a fare certainly not exceeding a halfpenny a mile, and the country-loving citizen will be able to journey at the same outlay to faraway centres of sylvan scenery.'

The *Tramway & Railway World* in its July 11th issue considered the inauguration one of the most important events in British electric traction history. In transplanting the overcrowded masses to country or suburbs the LUT appealed to philosophers and politicians and to all concerned in the housing question. The great men who attended the ceremony recognized a new factor in civilization. The company's policy was to lay lines and await the traffic. On the inaugural trip to Southall, said the journal, it was noticeable that many houses were being erected along the route.

Hardly had the Southall service got going before complaints of noise began to fill the local papers. The Rev. Henry C. Douglass, of St Matthews, Ealing Common, waxed particularly hot. 'The ear-splitting thunder and nerve-scraping resonance of Mr Clifton Robinson's gaudy tumbrils. . . . Each car is rapidly developing a sonorousness of its own. There are some now that emit a giant sigh and a monster pant, and even whistle and scream in addition to the general rumble and rattle and grind. . . . We welcome the tramway as a cheap, quick, convenient and pleasant method of travelling. . . . But this noise we cannot endure.'

Robinson retorted that one could not please everybody. Some people had even said that the cars gave insufficient notice of their approach. 'Were our cars absolutely noiseless, Mr Douglass would still find some excuse to advertise his resemblance to Ruskin and to indulge his passion for strong language.' He was backed up by a contributor who said it was a case of getting used to an unaccustomed quality of noise. Another writer thought trams a really valuable kind of locomotion, though noise was a serious objection.

Robinson reserved his chief scorn for a writer who thought that the white cars were noisier than others as they seemed to have been built in Pittsburgh – and were, therefore, American rattletraps! (The writer was misled by the wording on the controllers, which were from American Westinghouse.) The red cars explained Robinson, were from Hurst Nelson. The white cars cost more – more than £1,000 apiece – and came from Milnes. They were, in the opinion of qualified judges, the *ne plus ultra* of electric tramcars, 'built by the oldest established and most competent firm of tramway car builders in this or any other country. I designed these cars myself and am a little proud of the fact that several large towns which are going in for electric tramways are eager to obtain permission to adopt every detail of my designs'.

In fact, the white cars proved the least successful of the fleet, partly, it is said, because their McGuire trucks overhung more than other types and so tended to strike other vehicles on curves.

Junkets – And Hard Bargains

A Colonel St Maur-Wynch of the Indian Army declined to have anything to do with a circular got up in Ealing to complain of the noise. 'Though I sleep in the front room of the house and keep my window open at night, the running of the tramcars does not either inconvenience or disturb me; but then my liver is in good working order.' He hoped that when he returned to Ealing from India in a year's time the projected Ealing-Brentford line would be operating.

Ealing Corporation then began a long battle with the LUT over noise. It won an early round when interlaced track was replaced by double track at the Hanwell boundary, complained of as being a noisy spot.

The LGOC was too old a hand to be caught completely napping by the LUT. It spruced up its horse buses on the Oxford Circus–Hanwell and Tottenham Court Road–Ealing routes and put on a new Acton–Charing Cross service. It could not beat the LUT for speed. Instead it cut fares to 4d throughout, with 2d between Hanwell and Shepherds Bush, for which the LUT charged 3d. This fare reduction also hit the Central London.

Meanwhile, the company's ambitious Bill in the 1901 Session was taking its course. It covered no fewer than $17\frac{1}{2}$ miles of tramway in the County of London and extensive tramways in Middlesex and Surrey, including the Kingston area lines already detailed. The Middlesex and Surrey proposals, including new light railways for which Orders were sought, covered $27\frac{1}{2}$ miles of line.

The inner London proposals were as follows:

Hammersmith Broadway–Willesden (Harrow Road), via Shepherds Bush and Wood Lane

Shepherds Bush–Camden Town, via Notting Hill, Royal Oak, and St Johns Wood

Harrow Road (junction with Shepherds Bush–Camden Town line)–Marble Arch

Hammersmith Broadway–Hyde Park Corner, via Kensington High Street

Holland Road (linking Shepherds Bush with Hammersmith–Kensington line).

Hammersmith Broadway–Putney Bridge via Fulham Palace Road (a curtailment of the original Hammersmith–Putney–Kingston proposal) Hammersmith Broadway over Hammersmith Bridge.

The other tramway proposals were:

Hammersmith Bridge–Barnes Common
Richmond (Kew Road)–East Sheen–Barnes Common–Putney (Upper Richmond Road)
Brentford (High Street)–Ealing (Castle Bar Road/Longfield Road), via Ealing Road, St Mary's Road, Ealing High Street, Spring Bridge Road, Haven Green and Castle Bar Road.
Teddington (Broad Street)–Hampton Hill
Isleworth–Richmond Bridge via St Margarets (with Crown Road spur)
Hounslow–Cranford

Light railway proposals included:

Hounslow–Southall via Heston and Frogmore Green
Hampton–Sunbury Cross via Kempton Park
Acton–Willesden via Horn Lane and Old Oak Lane
Hounslow–Hanworth
Twickenham–Hanworth–Sunbury Cross
'Askew Arms'–Old Oak Lane via Friars Lane.

The inner London routes were almost entirely double track. The Bill included powers to widen many streets and roads, including, within the County of London, Bramley, Silchester, Warwick and Harrow roads, Maida Hill West, St Johns Wood Road, Albert Road, Fulham Palace Road, Fulham High Street and Upper Richmond Road.

It is particularly interesting to speculate on a Shepherds Bush–Camden Town link, which would have placed the LUT in direct communication with populous north and east London and afforded fascinating opportunities for through journeys, if not for through running. But, alas, no LUT short workings would ever carry the destination 'Zoological Gardens', nor would the denizens of West London be able to shop at Barkers, Derry & Toms or Pontings by tram. When the Bill came before the Commons the Standing Orders Committee struck out all lines lying wholly within the County of London, as the LCC had uncompromisingly vetoed the lot.

At least the Committee spared the Hammersmith Broadway–Bridge and the Putney part of the Richmond–Putney lines as they formed part of through routes of which more than two-thirds lay in Surrey. Even these exemptions were disallowed when Sir John Kennaway's Committee got to grips with the Bill.

Lord Kinnoul headed a West End Tramways Opposition Association,

Junkets – And Hard Bargains

declaring that 'the introduction of tramways almost invariably brings about great depreciation in the value of residential and other property as well as many other disadvantages'. Robinson telegraphed towns with tramways, all of which refuted the allegation, but Kensington and Paddington were not to be moved.

Hammersmith remained an ally and deplored the LCC's attempt to pose as a tramway promoter mainly to relieve the housing problem while it tried to block the LUT's attempts to do the same thing.

When the Bill came before the Commons Committee, Balfour Browne was able to quote some remarkable traffic figures which he considered unparalleled. On the 7 miles 4 furlongs so far opened, 5,480,208 passengers – almost the population of London – had been carried. The daily average was 71,260. On Whit Monday they had carried 210,557.

The LUT had agreed to spend large sums at the behest of local authorities. In Surrey alone the total was £1,000,000. Even so, when, recently, the company had put £350,000 of debentures on the market at 4 per cent they had been over-subscribed five or six times. To object to electric tramways, averred Balfour Browne, was to be behind the times. If individuals suffered slightly the public gained enormous advantages.

It was stated that in view of the high capital cost of the electric lines, the local authorities had agreed to extend the purchase period from twenty-one to twenty-five years on consideration of annual payments by the company – a doubtful bargain.

James Bigwood, as Chairman of Middlesex County Council's Parliamentary Committee, thought that the LUT lines would sooner or later come into the hands of the county.

E. J. Halsey, speaking for Surrey, said that he would not be surprised if the LUT did not one day link with London over Kingston Hill as there was a great want in east–west Surrey communication. It was necessary, Halsey considered, to widen Kingston Bridge. The LUT offer of £10,000 for that was fair and its further offer of another £10,000 for Hampton Court Bridge would suffice for a widening that would serve for some time. [Hampton Court Bridge, with a carriageway some 21 feet wide, lay slightly north of the present structure and was aligned with Bridge Road, East Molesey.]

Robinson said that there would be no difficulty in widening Kingston Bridge, which was 16 feet 9 inches wide. The present width would still allow 1 foot 10 inches between the base of tramcars and the kerb. He did not agree that trams would add to the heavy traffic using the bridge.

London United Tramways

By July 1st, Robinson declared, he had 150 cars for the 16 miles built. When the present lines were ready he expected to carry 52,000,000 passengers a year. In the past ten weeks they had carried 100,000 workmen at a ½d a mile. He feared to state the total of passengers when all their routes were open. In summer they would run a car to Hampton Court with 70 passengers every minute of 20 hours. They were educating people to ride instead of walk. Already, at Kew Bridge, they had unloaded 70 passengers from a car and started another with 70 passengers every 45 seconds for 20 hours.

They would hold their hands in respect of Richmond Bridge – 'a very beautiful bridge indeed' – with a view some day of throwing a new bridge for trams across the river close to the railway bridge. They would do so when Richmond Council wished, probably when the new Kew Bridge was completed.

Their proposals included an extension of time for the Boston Road line as they did not think they could get it ready within the authorized period.

Residents of Castlenau, (Barnes) and members of the Ranelagh Club, were against a tramway along Castlenau. Sir Lewis McIver, MP, a member of the club, said a tramway would destroy the approach to the club from Hammersmith Bridge and spoil it for team driving. This section was an essential part of the proposed T-shaped route from Hammersmith to Barnes Common, Putney and Richmond, there linking with the Kew Road line which it was proposed to convert. In respect of this group of lines, the LUT was prepared to spend on widenings £28,797 in Richmond, £15,143 in Barnes, £43,618 in Wandsworth and £900 in Hammersmith.

Robinson said he was aware of the memorial got up by the clients of Lord Robert Cecil, representing frontagers in Sheen Road and elsewhere. The 'game' had begun at a public house in 'Richmond Road' and was engineered by a publican and a bus company.

The cyclists who used 'Chiswick Road' and 'Richmond Road' were like mosquitos.

> Cecil: And you kill them like mosquitos. How many bicyclists have you bagged?
> Robinson: Accidents will happen in the best regulated families, and particularly on electric tramways.
> Cecil: Oh! on electric tramways.

Robinson: Yes, because it's a novelty. People are not educated up to the electric tramway, particularly in densely populated districts.'

Robinson said that many widenings were contemplated in the Bill. They could not all be done without future powers but many could be done by agreement. They had dropped lines that local authorities had persistently opposed.

Robinson hoped that Standing Orders could be amended to dispense with the imperative requirements for consent. He considered it audacious of Middlesex council to seek an order allowing them to run over the LUT.

The proposed extension from Hounslow ('Bell') along Bath Road to Cranford was included at the request of the Heston-Isleworth Council. It would be of great benefit to Hounslow Barracks. But it clashed with new light railway proposals of Middlesex which will be considered shortly.

Cecil was successful in inducing the Committee to throw out the 'Barnes' proposals. The Kew Road conversion was passed, but the Bath Road line was referred to the Light Railway Commissioners, who eventually sanctioned it. The Kingston group also passed, including the crossing of both Kingston and Hampton Court Bridges, the chairman stating that the widening of Kingston Bridge should coincide with the tracklaying. The original intention had been to approach Surbiton by way of Surbiton and Maple roads to Claremont Road, but to satisfy the owner of Surbiton Hall, the route was amended to traverse Surbiton Crescent instead.

In July, the Lords Committee struck from the Bill the clauses relating to 25-year purchase by Surrey on the grounds that the county council was not the tramway authority. It also struck out the Hammersmith-Richmond line. At this, Richmond Council, which had come round to backing the conversion of the Kew Road line, turned about and again opposed, as that line seemed destined to remain isolated. All the same the council would electrify the line if it possessed it.

The Middlesex County Council proposals now included the following lines:

Staines-Hampton (to join authorized LUT line)
Staines-Hounslow Barracks Station (on right-of-way east of Baber Bridge, Hounslow Heath)
Hounslow ('Bell')-Cranford-Colnbrook

London United Tramways

Hounslow–Southall, with spur to the 'Grand Junction'
Hounslow–Hanworth
Twickenham–Hanworth–Sunbury Cross.

The intention was to build the lines and then lease them for operation to the Metropolitan Tramways & Omnibus Co. Ltd; a company formed in 1894 to build and operate electric tramways in Middlesex and Hertfordshire. But the MT & O, which became the Metropolitan Electric Tramways Ltd on acquisition by the BET group in November 1901, did not consider that lines in west Middlesex would pay enough for some time and was not prepared to come in. (In fact the MET later operated all the routes which the council subsequently built in the north-western and northern suburbs.)

Middlesex had been in touch with the District Railway and hoped there might be joint working, with Middlesex cars possibly running on to the District's Hounslow line at Hounslow Barracks.

Charles Tyson Yerkes, of Chicago, representing powerful American syndicates, had just secured control of the District. Hardly had he formed the Metropolitan District Electric Traction Co. Ltd, to take over the control of the District and electrify it, when the Light Railway Commissioners began their inquiry into the light railway schemes of the LUT and Middlesex Council.

Mr Lewis Coward, KC, for the LUT, asked Yerkes when he took over the lines (the District Railway). 'Yerkes: Yesterday (laughter). Mr Coward: Oh, only yesterday – then I will ask you no more (renewed laughter).'

Mr H. H. Asquith, KC, MP, (later Lord Oxford and Asquith), for Middlesex, stressed the fact that the council would widen to 50 feet all the roads which their light railways would traverse. He said that there were many accidents in narrow Brentford High Street. The light railways would feed rather than compete with the LSWR.

Alderman Bigwood rather spoilt the last point by recounting that he had recently missed a train at Waterloo, had doubled back to the Bank, gone by Central London tube to Shepherds Bush, then by LUT to Hounslow and had still kept an appointment there (laughter and applause from the LUT side).

Counsel for the LSWR considered that they might lose 30 per cent of their Richmond–Staines traffic if a light railway linked those towns.

When the inquiry turned on the question of attracting the lower classes to live in places like Longford, Bigwood related that a deputation had once come to him from Richmond 'with their gold chains and so on' and had said: 'We don't want you to widen Kew Bridge for the tramways because they will bring over to us all your rabble from Brentford.' But when they found there was a chance that they might possess tramways themselves they said: 'We shall be very glad to have your bridge made wider and to have your people from Brentford come over to visit us.'

Southall–Norwood council's terms for the Southall–Hounslow light railway were exacting in the extreme, in contrast to the way in which it had welcomed the Uxbridge line. It required the LUT to widen the roads to 50 feet throughout and pave with hardwood throughout. Only 100 yards could be broken up at any one time. The company was to erect at least one waiting room, with free conveniences. Cars were not to wait anywhere longer than two minutes. All dirt was to be removed from the rail grooves by wire brushes, save only when the whole road was being watered. The fare between Southall High Street and Heston church was to be 1d.

The LUT proposed to lay double track throughout and form junctions at Southall and with the authorized Baber Bridge and proposed Cranford extensions, in Hounslow. The estimated cost of construction was £57,500 but widening, hardwood paving and reinstatement of the road surface took another £70,500! And all this for a mere 3 miles 7 furlongs 5 chains of country road!

It was a similar story with the Sunbury–Hampton line. Construction was to cost £40,000 and widenings (to the 50 feet demanded by Middlesex), and other works another £44,000, for 2 miles 6 furlongs of route.

The Light Railway Commissioners, who had approved the 2-mile 7-chain Cranford extension as a light railway on July 29, 1901, opened their inquiry into the LUT's light railway proposals on March 19, 1902. Excluded already were 2 miles 4 furlongs of the Acton–Willesden via Old Oak line. Although this lay within the agreed LUT sphere of influence, Robinson withdrew the scheme in deference to Middlesex, which was to take seven years to see the line through – it was opened on October 7, 1909. The associated link from 'Askew Arms' was vetoed.

The Hounslow–Hanworth and Twickenham–Hanworth light railways were sanctioned by the Commissioners, subject to Board of Trade approval. Unfortunately the lines were never built. But the exorbitant demands in respect of the Southall–Hounslow, Hanworth–Sunbury and Hampton–Sunbury schemes forced the LUT to withdraw them.

Nevertheless, the Act of August 17, 1901, though extremely hard won, meant that the LUT would gain a firm footing in Surrey, where the county council, at least, was more co-operative.

As for the tramway proposals, Standing Orders had been allowed to be dispensed with and the Bill allowed to proceed, but it was pruned of all its London offshoots and the local veto accounted for the little line in Teddington and the St Margaret's link.

Chiswick depot was the scene of more festivities on August 6, 1901, when young Clifton Robinson was entertained to a coming-of-age supper. More than 1,250 invitations to every LUT man and boy were issued and all responded.

One of the car sheds had been transformed into a supper room. Whiteleys catered and the string band of T Division of the Metropolitan Police played. Tables groaned under galantine, roast beef, pies, pressed beef, ham, salad, custard puddings and tartlets. There were beer, ginger beer, lemonade, tea, coffee and tobacco. Guests included W. J. Bull and Reuben Cramp Rogers, superintendent of the Teesside tramways and related to the Robinsons.

No young man, supposed W. J. Bull, had ever begun life with fairer prospects than Clifton Robinson junior. He was the son of a rich man who had not been so foolish as to bring him up to a life of idleness or self-indulgence. He had health, energy and good looks and he held a remarkable position for a man of twenty-one. He fulfilled responsible duties in a way which had won him the love and esteem of all with whom he was associated.

In his reply, the subject of this eulogy – who was presented with a diamond ring and an illuminated address on behalf of the LUT staff, and a pearl scarf pin – said it would ever be his desire to follow in his father's footsteps. 'Let my career be what it may, I can promise you one and all that your kindness and encouragement to me, on the threshold of that career, shall always be remembered with kindness, consideration and esteem.'

Clifton Robinson hoped that the memory of that night would stimulate his son, in the years when perhaps he himself had gone, to carry on the work on which they were all engaged. Heavens knew where their ramifications would extend. When all their lines were in operation 'I want to see you all round me with your numbers augmented a hundredfold'. No man had lost his position because of the electrification. 'We honour our chairman, we serve the public and we trust in God.'

One account of the junket described the scene which followed the toasts, with 'hundreds of the men rushing from their tables into the broad central space . . . the great hall was all a-flutter with waving hats and handkerchiefs, the white-crowned yachting caps of the operating staff adding a great touch of singularity to the sea of heads'.

The *Middlesex County Times* recalled that at an age when most young men of his class were playing boyish pranks at Oxford or Cambridge, young Mr Robinson had settled down to the stern business of life. He had been apprenticed to his father from birth, and had received a tramway training such as no other youth ever had.

Indeed, since the Board had elected him to be traffic superintendent in 1899, young Robinson had come to have absolute control of 1,000 men. All traffic matters were his to settle. He kept discipline while keeping the esteem and regard of the staff by his good temper and cheery manner.

It is said that on finding a gang on the Hounslow line grievously at fault, he sacked every man. Later that day the men came to Chiswick to beg for reinstatement. Young Robinson wanted neither to lose nor spoil them. 'Come back in two hours,' he said, 'and I'll tell you what I'll do with you.' The suspense had the desired effect and that gang gave no more trouble.

In June 1901 more than 100 officials, drivers and conductors from Chiswick had enjoyed a day's outing to Cobham. The 'young boss' had driven down to join them at the 'Royal Oak'. His father, busy at Westminster, had wired: 'Trust they will thoroughly enjoy themselves and lay in a stock of health and energy for the approaching busy time.' After lunch there had been sports and a singing contest. For the Acton depot staff there had been a day out at Marlow.

LUT outdoor staff had two outfits of uniform a year. Drivers and conductors then worked two shifts per car, about ten hours a day. The men ran their own benevolent fund, the directors doubling the total subscription. The fund met medical costs and paid half-wages to those sick.

The company reinstated reservists back from the South African War. While men were with the colours, their wives received 10s 6d a week, plus 1s for each child.

In September 1901 Colonel R. E. Crompton addressed the British Association on street traffic. He proposed new main roads traversed by motorbuses running at 16 m.p.h. Robinson felt obliged to refute some of his statements and figures in a letter to *The Times*. The LUT, he said, had often carried 7,500 passengers an hour in one direction past Young's Corner. Their 120 cars could move 8,400 passengers in an hour. As existing ordinary buses held only 26, Crompton would need almost 550 to carry his total of 14,000 passengers an hour. Even to reach 8 m.p.h. he would have to use them on a road reserved for buses. At Mansion House, and some other places, 500 buses passed an hour, but at an average of 1 rather than 8 m.p.h.

The future, hazarded Robinson, might well be with motor traction on special roads. It might even be with flying machines. For the present, in London, it was with the tram – the tram plus the tube. Robinson also foreshadowed the transport co-ordination of London, a theme which he was soon to elaborate and, at least in part, apply.

In its issue of October 17, 1901, the *Tramway & Railway World* asked in connection with LUT progress: 'Are we going to have the beginning of the end of overcrowding now?'

Although many of the lines authorized were along quiet rural roads, there was the hopeful example of Southall, where large numbers of working class houses were planned.

Riding on the top deck had become a great diversion and an amusing article in a West London paper pointed out 'the fascinating new domestic vistas possible – 'it is infinitely more interesting to know your neighbours'.

Chiswick Council considered that the LUT, preoccupied with grandiose outer extensions, had overlooked its needs. Chiswick was fast growing, with 30,000 inhabitants, more than three times as many as thirty years previously.

In 1900 the LUT had projected a series of local lines in Chiswick and Acton. One of the routes would have traversed Gunnersbury Lane, Bollo Lane, Bollo Bridge Road, Acton Lane, South Parade, Turnham Green Terrace, Annandale Road, Devonshire Road, Mawson Lane and Chiswick Lane.

About July 1901 both the LUT and Middlesex County Council sought Light Railway Orders for rival schemes in the district. Chiswick Council would have stood by its agreement with the LUT but was stung by the threat of Middlesex intervention to promote its own Bill for tramway powers in the 1902 Session. The *Chiswick Times* said that a 'toy scheme' would be costly and inefficient but believed a joint LUT–Chiswick project might work.

In the event the council withdrew its Bill in April 1902 and both the

Junkets – And Hard Bargains

LUT and Middlesex applications were thereupon also withdrawn. But, as will be shown, Chiswick still cherished aspirations for a local system.

The year 1901 ended with news of a projected Hounslow, Slough & Datchet Light Railway, promoted by the new Metropolitan District Electric Traction Company, to run from Hounslow Barracks station along the Bath Road to Slough, with a branch from Colnbrook to Datchet. It was to be laid almost entirely on main roads but the initial section at Hounslow was to be on right-of-way and a new road was planned at Colnbrook to avoid the Great Western Railway level crossing.

Chapter 6

New Ground – And 'Underground'

A complicated situation had developed in the Wimbledon area. After consultation with the Croydon Rural District Council the Wimbledon Urban District Council had decided, late in 1900, to promote a Bill for a tramway between Wimbledon Broadway and the county boundary at Longley Road, Tooting, there to join a possible extension of the LCC tramways from Upper Tooting.

In November 1900 Robinson asked that when Wimbledon Council gained its powers it should allow the LUT to compete with any other undertakings invited to lease the line.

Wimbledon estimated the cost of the line at £32,689 for 1.97 miles with overhead, or £57,289 – reckoning the cost of four cars at £850 each – if the council worked the line. In July 1901, after it had been agreed to apply for a Provisional Order in the next Session, the surveyor was asked to estimate on the basis of a larger system with lines from the county boundary on Wimbledon Common to Merton High Street, along Quicks and Haydons roads to Plough Lane and the county boundary at Summerstown, and along Worple Road to Raynes Park.

In August 1901 Robinson sought to negotiate with Wimbledon, now that his Kingston lines had been sanctioned. At the same time, instigated by Mitcham Parish Council, Croydon RDC had applied to build light railways from the LCC boundary at Tooting Junction to Mitcham and thence to the Croydon borough boundary. In its first form the Mitcham plan proposed lines between Tooting Junction, Mitcham and Carshalton, Mitcham and Croydon, and Tooting (Longley Road) and Merton.

Before the Mitcham Order was confirmed by the Board of Trade both the LUT and BET had sought to obtain it. In the event the BET was successful, the Mitcham lines eventually passing to the BET-controlled South Metropolitan Electric Tramways & Lighting Co. Ltd, formed in 1904. But the BET was not so fortunate in Wimbledon, where it found the council's terms for lines linking Wimbledon, Merton and Summerstown quite unacceptable.

Robinson was prepared to swallow them. Wimbledon was vital to his design to link the Surrey lines with South London and make through running to central London possible.

The main conditions were as follows. Double track was to be laid in Worple Road. Wimbledon Hill Road was to be widened to 50 feet near the LSWR station bridge. The LUT was to foot the whole bill for road widenings, not merely £30,000 as it had offered. There was to be no extension up Wimbledon Hill and alongside Wimbledon Common without special consent and even then the LUT would have to pay £5,000 toward the cost of widening High Street at Church Road junction before laying track there.

Widenings were to allow for 8-foot-wide footpaths, except by consent. An annual rental of £100 a mile was exacted. All roads traversed by the tramways were to be wood paved. There were to be no advertisements displayed on trams running through the town.

A map published in October 1901 showed a huge crop of projects for which the LUT intended to seek powers in the next Session. In addition to an extension from Malden to Raynes Park, Wimbledon, Merton, and Tooting boundary, a Merton–Summerstown branch, and a line from Hammersmith Broadway over Hammersmith Bridge, along Lonsdale Road to Barnes riverside and thence along Lower Mortlake Road to Richmond, there were the following proposals:

Kingston Hill ('George')–Kingston Vale (county boundary)
Wimbledon Hill Road–Wimbledon Common (county boundary)
Merton (Grove)–Morden–Ewell–Epsom
Stonecot Hill (on Merton–Epsom line)–Sutton Station via Sutton Common Road and Sutton High Street
Colliers Wood (on Merton–Tooting section)–Mitcham via Church Road and thence to Carshalton and Wallington
Mitcham–Croydon (Surrey Street)
Tolworth ('Red Lion')–Ewell–Cheam–Sutton–Wallington–Croydon (Surrey Street)
Winters Bridge (Dittons)–Esher (Broom Hill)
Ham Boundary–Petersham

All the lines tabulated, save those to Kingston Vale and Wimbledon Common, were excluded by the time the Bill was submitted, partly because they involved much unremunerative rural construction and partly because of a pending territorial agreement with the BET. Such an agreement was reached in December 1901. The LUT was to 'restrict' itself to an area bounded by the Thames, Wimbledon, Sutton and Leatherhead, while the BET was to be free to exploit the 'hinterland'.

The outcome was particularly unfortunate in one respect. In the event the LUT never revived the Colliers Wood–Mitcham proposal – or any of the others – so that no direct link with the Croydon area was ever formed and what might have been a useful cross-country facility was lost.

It may be convenient here to dispose of a somewhat similar proposal which the BET put forward in both the 1902 and 1903 Sessions in its Croydon & District Electric Tramways Bill. It was for a link between the Longley Road terminus of the now authorized LCC extension, along Longley Road to join, by a hairpin bend, that of the authorized Mitcham Light Railway at Tooting Junction, permitting somewhat circuitous through running. It should be noted that in neither case was a southward curve to the LUT proposed at Longley Road/Tooting High Street.

The LUT Bill for the 1902 Session crystallized as follows: Shepherds Bush–Marble Arch
Hammersmith (Glenthorne Road)–'Askew Arms' via Dalling, Paddenswick and Askew Roads
Hammersmith Broadway–Hammersmith Bridge–Lonsdale Road–Barnes Terrace–High Street Mortlake–Lower Richmond Road–Lower Mortlake Road–Richmond
Doubling of Kew Road line, with extension to Richmond (Station Hotel)
Malden–Raynes Park–Worple Road–Wimbledon Hill Road–Wimbledon Broadway–Merton High Street–Colliers Wood High Street–Tooting (Longley Road)
Wimbledon loop (St Georges Road and Francis Grove)
Merton High Street–Haydons Road–Plough Lane–Summerstown ('Plough')
Wimbledon Hill Road–Wimbledon Common (county boundary)
Kingston Hill ('George')–Kingston Vale (county boundary)

The Marble Arch extension, after the summary rejection of inner London schemes in the preceding Session, seems an unnecessary tempting of fate. Perhaps Robinson was sentimentally attached to the idea of a line taking in the route of Train's tramway of forty years before!

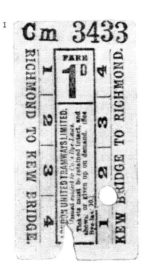

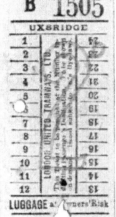

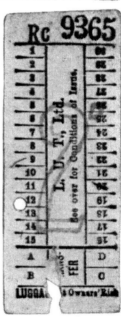

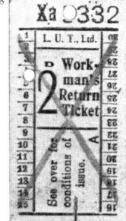
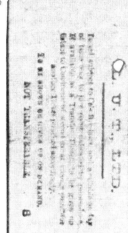
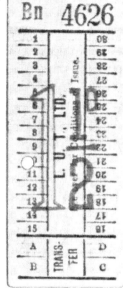
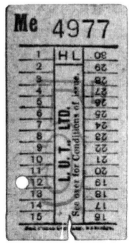

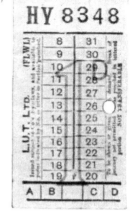
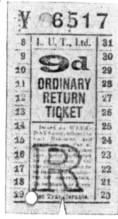

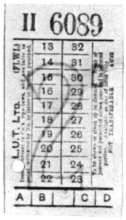

New Ground – And 'Underground'

The Bill included powers for much acquisition of land. It stipulated that purchase by Richmond and Barnes councils was to be made only in accordance with agreements with the company. The purchase period in respect of Wimbledon, Maldens & Coombe, Merton and Mitcham was to be twenty-five instead of twenty-one years, at a fair market value.

A Commons committee began to consider the Bill on May 29, 1902. Balfour Browne said that the LCC and LUT had reached an understanding on constructing a line over Hammersmith Bridge, a suspension bridge opened in 1887, with a 27-foot 2-inch carriageway narrowing to 19 feet 9 inches between the towers.

For the first time Richmond's attitude seemed to be straightforward. At least it did not oppose, a forbearance explained perhaps by the fact that the LUT had agreed to meet a huge bill for widenings. (The amount for widenings and other works in Richmond, Barnes and Hammersmith was £158,752.) Yet another £30,000 was to be spent on forming a riverside terrace at Barnes which would also prevent flooding.

To its credit, Barnes Council offered to grant Robinson a fifty-year purchase period in exchange for the esplanade improvement but the Committee chairman Compton-Rickett reduced the term to thirty years.

Widenings in Wimbledon alone would cost the company £180,000. It says much for the faith of Robinson and his fellow directors that they could even consider such outlay before new lines brought in a penny of revenue. They certainly gambled hard on a rise in population in the areas they served, and on increasing patronage. Robinson told the committee that at the last census their 16 miles of line so far open served a population of 412,000, which was rapidly increasing.

The veto of frontagers was enough to induce the committee to disallow the Kingston Vale and Wimbledon Common extensions, but the LUT would be allowed to submit evidence in support of a shorter extension up Wimbledon Hill alone. Robinson said that a line up the hill would not injure property. Cars on that section would be limited to 10 m.p.h. The chairman of the Wimbledon & Putney Commons Conservators considered that such a line would not only impair amenities but also add to traffic hazards on the hill.

The LUT proposed a 5d fare between Shepherds Bush and Richmond via Hammersmith and Barnes, compared with the LSWR's 1s 3d third class fare. Robinson told the committee that they carried workmen between Southall and Shepherds Bush for 1d between 4 and 7 a.m. They were building new lines at a rate of about 1½ miles a month and paying out £10,000 a week in wages.

The LSWR withdrew opposition to the Bill on engineering grounds but continued to oppose on competition. Surrey County Council backed it. Halsey said that Surrey had resisted an attempt by the LCC to become a tramway authority in Surrey. They had considered becoming a tramway authority themselves but had never achieved the required two-thirds majority.

The noted engineer Sir Douglas Fox, himself a Surrey alderman, gave evidence in support of the Bill. He considered that Hammersmith Bridge, although a suspension bridge, was strong enough for tramcars, but the decking might require cross-bracing to prevent oscillation. He advocated restricting the number of cars crossing at any one time.

The LCC had changed its views on a tramway across the bridge in spite of favourable engineering evidence and the fact that the LUT undertook to end the line 300 yards short of the Broadway.

The LUT agreed to widen Merton High Street and pay £100 a year per mile of track in Surrey. It further declared its willingness to extend 300 yards at the Longley Road terminus to form an end-on connection with the authorized Tooting extension of the LCC.

Allies came forward, at a price. Councillor Thompson, of Richmond Tramways Committee, and the Chairman of Barnes Highway Committee supported the Bill, and the Thames Conservancy approved of the proposed embankment at Barnes.

While Parliament debated, the construction gangs were working all out on the authorized lines. Large squads toiled between Richmond Bridge and Twickenham, where work went on day and night, and between Hampton Wick and Hampton Court.

Work in Teddington was held up pending the end of negotiations with the council, which was being difficult. The LUT threatened arbitration by the Board of Trade if the council persisted in refusing to pass plans of the already authorized lines.

Difficulty of another kind was experienced in Heath Road, Twickenham, where the mains were found to be close to the road surface. The council agreed to levels being raised if the frontagers affected agreed. By the middle of April, all the track and most of the standards were in place on the Richmond Bridge line, but some widening remained to be done. Mains also had to be lowered in Hampton.

New Ground – And 'Underground'

By the end of June trial running was possible between Busch Corner and Cole Park Road, Twickenham.

In March the LUT and Richmond had sealed their agreement. The LUT also offered £20,000 towards the cost of securing a right-of-way via the Quadrant, Eton Street, Paradise Road, Ormond Road and Red Lion Street. But by May the council had rescinded the agreement. Heston-Isleworth approved the Baber Bridge extension in May.

During the final stages of the hearing of the Bill, objections were heard from frontagers in Wimbledon and Barnes. Their opposition was reinforced by evidence from an Ealing surveyor who declared that rents of larger houses there had dropped since the trams had started running.

C. J. Owens, general manager of the LSWR, said that his company had carried 838,819 passengers in 1901 from Wimbledon and Raynes Park to stations on the route of competition, and had received £9,479. The Malden–Wimbledon tramway would almost duplicate the railway and might rob it of £6,000 revenue.

The LSWR's counsel said that tramway competition was losing the company £10,000 a year. Between Kew Bridge and Hounslow the gross receipts had dropped in six months from £1,516 to £524. The LSWR was spending £1,700,000 on improvements between Waterloo and Clapham Junction to improve traffic handling.

The LUT did pretty well with the Bill. Apart from the vetoed sections mentioned, the main loss was the section between Hammersmith Broadway and the south side of Hammersmith Bridge. At least the LCC now promised to seek powers in the next Session to rebuild the bridge and lay a tramway over it even though the existing structure was only fifteen years old and had cost £89,000.

The Act, passed on August 8, 1902, authorized the following new lines:

Malden–Wimbledon–Tooting (with short spur in Wimbledon Hill Road, and St George's Road loop)
Merton–Summerstown
Hammersmith Bridge (south side)–Barnes–Richmond
Askew Road line
Kew Road doubling

The total cost for 12 miles 6 furlongs 9.8 chains was £272,000. The Barnes embankment and terrace was now estimated to cost £40,000.

In addition the following wayleaves were payable:

Merton and Mitcham	£100 a mile each

Richmond	£1,250 a year
Wimbledon	£100 a mile
Maldens & Coombe	£150 a year
Surrey County Council	£100 a mile on main roads

Supplementary street and other works in connection with already authorized tramways would cost another £53,000.

In May 1902 the LUT applied for an Order for an ambitious light railway from Cranford to Colnbrook, Slough, Taplow and over Maidenhead Bridge to Maidenhead ('The Bear'), with branches to Datchet (Manor Hotel) and Langley ('North Star'). As with the Metropolitan District Electric Traction Company's project, a new by-pass road at Colnbrook was contemplated. Opposition by the Great Western Railway compelled the Light Railway Commissioners to reject it, on October 22, 1902, and direct that it should go to Parliament.

The extension beyond Hounslow ('Bell') was opened on Wednesday August 13, 1902, but only as far as the 'Hussar' (Barrack Road), short of the authorized Baber Bridge terminus. The Busch Corner–Cole's Bridge section opened the same day.

The Twickenham–Richmond Bridge and Cole's Bridge–Twickenham sections opened on September 13th.

In spite of all extortions and exactions things were going well. At the statutory meeting in early March 1902 Samuel White said that they had received so many applications for preference and debenture shares that on most large allocations the directors could allot only one-twentieth of the amounts asked for. Public confidence was not misplaced. The traffic figures were most encouraging. On Easter Monday, for example, 261,418 passengers were carried compared with 164,390 on Easter Monday 1901.

It did not seem that Robinson would succeed in his attempt to bring the London United into the West End. But a difference which arose with the Central London Railway seemed to offer another kind of approach. The Central London was planning to turn itself from a point-to-point into a circular line by extending from Shepherds Bush to Hammersmith, Kensington, Piccadilly, Strand, Fleet Street and so back to the City.

Robinson sensed a threat to LUT interests and aspirations. An interview with the CLR general manager proved unsatisfactory. Robinson told George White: 'We shall have to go in for tubes ourselves.' 'What do we know about tubes?' asked White. 'What we don't know we can learn', replied Robinson.

New Ground – And 'Underground'

The upshot was that a Bill was deposited in November 1901 to incorporate the London United Electric Railways Company, build railways, make agreements with the London United Tramways and with others, and acquire lands in Fulham and St George's for generating stations. The estimated cost of the projects was £4,413,047.

The tube lines proposed were from Barnes (Lowther Road) to Charing Cross Station via Hammersmith, Kensington, Hyde Park Corner, Constitution Hill, The Mall and Strand; Park Street (Marble Arch) to Clapham Junction (near Wandsworth Common) via Sloane Street, Queen's (now Queenstown) Road and Lavender Hill; Hammersmith–Brook Green Road–Shepherds Bush–Holland Road–Kensington Road (forming a loop of the first-mentioned line).

The LUER was one of twenty-six tube railway proposals which came before Parliament in the early part of 1902. The boom almost became a mania, with several companies competing, as in the Railway Mania of the 1840s, for the same sections of route. The Piccadilly–Strand route was one of the most coveted prizes. (A joint Select Committee on underground railways in 1901 had recommended that there should be a through tube railway between Hammersmith and the City via Piccadilly.)

So great was the pressure of Bills that a joint Parliamentary Committee, with Lord Windsor as president, sat from April to June to consider them.

A few days after Lord Windsor's committee had begun its sittings it was announced that the promoters of the Piccadilly & City Bill had reached agreement with those of the North East London (City–Tottenham–Palmers Green) and City & North East Suburban (City–Walthamstow). The C & NES was rejected as regards its Aldersgate–Mansion House section and the whole scheme was then withdrawn. The P & C and the NEL interests merged and presented one Bill.

In the midst of the fray entered the famous American banker, John Pierpont Morgan, keen to secure large interests in Britain. In May he announced that he had formed a syndicate to build about forty miles of tube railway, including the Piccadilly, City & North East London.

Robinson told the committee that the LUER would have no difficulty in providing an all-night service. The 13-foot 6-inch tunnels would be large enough to allow repairs and maintenance to be carried out while trains ran.

The line would run between what Robinson called the Scylla of the Albert Hall and the Charybdis of the Albert Memorial. But the centre line would be 150 feet from the hall and 75 feet down in the clay. They could not give the Albert Hall authorities the 'Bank of England' clause by which they would have to compensate them for any damage or injury – the LUER could never raise capital with such a condition.

The LUER and the Morgan group soon came to an agreement whereby the LUER would build between Hammersmith and Hyde Park Corner only and the Morgan group only the section from Hyde Park Corner eastwards, of the PC & NEL. The two sections would form an end-on junction at Hyde Park Corner and the line would be worked jointly. The LUER would abandon the Hyde Park Corner–Charing Cross section (which included a balloon loop) and the PC & NEL its own Hyde Park Corner–Hammersmith proposal.

In view of the agreement, which seemed to promise a strong new element in London transport, the Lords passed the combined scheme. But the Commons Committee was instructed to take security for the completion of the scheme under the joint arrangement.

Of the capital of the combined scheme, £11,380,000, Morgan was responsible for £7,690,000 and the LUER £3,690,000, in spite of the disparity of each partner's mileage. In any case the LUER rightly expected to have equal managerial control on the grounds of the well-established London United Tramway business and the traffic which the LUT could bring to the joint tube.

The alliance seems to have been uneasy from the start. The Morgan group, apparently, delayed ratifying the agreement on control which the LUER alleged had been made early in the proceedings, and indeed denied that there had been such an agreement.

One of the most important opponents of the Morgan–LUER project was the 'Yerkes' group, the Underground Electric Railways of London Ltd. Charles Tyson Yerkes, born in 1837, was an American financier and traction magnate who, after developing the street car system of Philadelphia in the 1870s, made a fortune with street cars and elevated railways in Chicago. At the turn of the century, like his compatriot Morgan and many others, Yerkes sought foreign fields for investment. He decided to turn his attention to London's transport and bought the powers of the Charing Cross, Euston & Hampstead tube, authorized in 1893 but unsuccessful in attracting investors. Aided by R. W. (later Sir Robert) Perks, solicitor for the transaction and a large District Railway stockholder, Yerkes gained control of the District in March 1901 and four months later incorporated the Metropolitan District Traction Co. Ltd to electrify it and build a power station at Lots Road, Chelsea. Perks became Chairman of the District on September 5, 1901.

New Ground – And 'Underground'

The UERL was a reconstitution, on April 9, 1902, of Yerkes' Metropolitan District Electric Traction Co. Ltd, of 1901, which by this time had not only control of the District Railway but had also acquired various companies possessing powers to build tubes. Two of the companies were the Brompton & Piccadilly Circus (South Kensington–Piccadilly Circus) and the Great Northern & Strand (Wood Green–Finsbury Park–Kings Cross–Holborn–Strand). Yerkes planned to merge the two, link them between Piccadilly Circus and Holborn, abandon north of Wood Green and extend westwards from South Kensington to Hammersmith by arrangement with the District.

By August 1902 relations between the LUER and Morgan were strained. Robinson resented what he considered the superior attitude of the Morgan group. Such lack of trust boded ill for a scheme which faced the powerful Yerkes interests.

According to Robinson, the Whites had decided at this time to sell their LUT shares and had entered into negotiations with various parties. Finally, they offered them to Speyer Bros. of London, one of the financial houses with which Yerkes was associated. Speyers bought them in the open market and so acquired a controlling interest in the LUT.

Another version of the story is that Robinson conferred with Perks and then went down to Cornwall to see White. Perhaps White then offered to sell his LUT holdings to enable Speyers to gain control. Certainly, when the deal was made known, the Morgan group alleged breach of faith and a secret arrangement with Speyers.

The ground was cut beneath the feet of Morgan. The new masters of the London United at once withdrew the LUER Bill. Parliament refused to sanction the truncated Morgan scheme on its own, as it no longer formed part of a through route. The Morgan interests attempted in vain to have the whole scheme recommitted as an amalgamation of the LUER and the PC & NEL, and Morgan thereupon withdrew from the London transit scene.

About this time Robinson is reported to have paid a flying visit to Ireland. He gave the hotel waiter a sovereign and asked him his name. 'Morgan,' replied the man. Robinson offered him a 'fiver' to change it!

Some time later a former Prime Minister is said to have remarked, 'Clifton Robinson? Wasn't he the man who got the better of Pierpont Morgan?'

Edgar (later Sir Edgar) Speyer told the Royal Commission on London Traffic in 1904 that the LUT shares had been offered to his firm after negotiations for their sale had been going on in another quarter. They availed themselves of the offer because the London United fitted in admirably with their railways.

So in September 1902 control of the London United passed to the Yerkes group and Yerkes himself took George White's place in the chair. He subsequently said that the LUT had formed 'the more natural alliance' with existing railway interests after months of endeavour had shown that the Morgan interests had no intention of concluding arrangements essential to protect the LUT. In two or three years a passenger would be able to enter an Underground Electric Railways Station almost anywhere in London and book through over the LUT, which would form a huge gathering ground for the Underground.

The LUT had been authorized to lay single track in the then narrow Hampton Court Road between Hampton and Hampton Court. But Robinson began negotiations with the Office of Works to obtain a strip of land of Bushy Park and Hampton Court Green to avoid the still more costly expedient of buying private property on the Thames side of the road.

An interesting by-product of the extension was the purchase of Garrick Villa, a fine Palladian House in Hampton, close to the corner of Church Street and Hampton Court Road. It was necessary to set back the high wall screening the house from Hampton Court Road. The company found itself having to buy the whole estate which included not only the house and large grounds at the rear but also its lawns on the river side, linked with the rest of the estate by a tunnel under the road and including a Grecian temple built to house a statue of Shakespeare, by Roubilliac. There was also a small island called Garrick's Ait.

The villa had originally been formed by merging some old cottages. David Garrick became its tenant in January 1745 and bought it the same summer. He altered and enlarged the house and commissioned Adam to design a new front. Dr Johnson was a visitor and it is at Garrick Villa that he remarked, 'Ah, David, it is the leaving of such places that makes a death-bed terrible.'

The tunnel under the road was built on Johnson's advice.

After Garrick's death in January 1779, the house was occupied by his widow for another forty-three years. When the London United came on the scene the owner was the widow of Edward Grove, a Lambeth clothier, who had made further enlargements. The company took its new acquisition seriously and was determined to maintain the amenities and associations. By its purchase, completed in November, it became probably the first tramway to own a fine historical mansion.

Part of the grounds were added to Bushy Park to compensate for the strip taken. Some 20 feet of the frontage was sacrificed to road widening.

The rest of the grounds and the house were thrown open to the public on regatta and fete days. A special siding was built close to the corner of Church Street – which had been widened at this point – to allow excursion trams, hireable by parties, to be stabled just inside the grounds. Subsequently, the siding housed the special single-deck car used by Robinson when in residence at Garrick Villa.

Street improvements on the Thames Valley extension cost £202,000 for the 11 miles 7 furlongs 7.6 chains. The setting back of the Bushy Park frontage and the construction of new ornamental railings and the widening at Garrick Villa took more than £30,000. Work in Heath Road, Twickenham, cost £30,000 alone. Some £5,000 went in lowering mains and widening under the railway bridge. The old brick bridge at Cole's Bridge was replaced by a new 45-foot-wide structure. Fulwell railway bridge was widened at a cost of £2,000. That at Teddington was rebuilt and widened at a cost of £7,000.

Chapter 7

Mainly Robinson

At the beginning of 1902 Heston-Isleworth Council sought powers to acquire land to widen Twickenham Road because of the alleged danger from tramway operation. The rails had been laid close to the kerb in some places and at others there was no footpath. Up to 800 trams – 1,000 in summer – used the route daily. The Council said Colonel Yorke had agreed that there were some danger spots but the LUT would do nothing about them.

At the LUT extraordinary meeting in February 1902 George White said it was desirable to increase the preferred capital from £825,000 to £1,000,000 and the ordinary capital likewise. All existing preferred and ordinary shares had been issued.

In March 1902 Robinson read a masterly paper on 'Electric Traction: London's Tubes, Trams and Trains' before the Society (now Royal Society) of Arts. He first briefly reviewed the chequered development of tramways in Britain. The 1870 Act might have had as its preamble: 'whereas it is desirable to discourage the construction of tramways in this country'. It made the promotion of electric tramways most arduous and in some respects impossible. Small wonder Britain lagged behind in electrical invention and original electrical development, so that it was necessary to go to the United States for much plant and machinery.

Nevertheless the LUT was carrying on its 16 miles of line more than one-twentieth of the total number of passengers carried on the 1,000 miles of British tramways in 1895. Mr St John Hankin was correct, if somewhat Hibernian, said Robinson, in arguing in the *Daily Chronicle* that 'it is the absence of tramways which crowd [*sic*] the London streets'.

By paying £10,000 to be rid of the observatory trouble the LUT had earned the thanks of every electric tramway or railway promoter, director and shareholder in London. The City & South London, Waterloo & City and Central London railways were lucky to have gained their powers before the claim arose.

Mainly Robinson

Tramways paid local rates in full, without the 75 per cent allowance conceded to railways. But even negotiations with three county councils and some thirty local councils did not daunt the LUT. The public saw only an operation as nearly perfect as anything human could be, and with extraordinarily low fares.

Robinson made the surprising admission that the cost – £30,000 – of installing overhead had been more than the ordinary estimate for conduit. But they had felt that overhead was better for outlying districts. He would prefer a contact system to any known type of conduit but circumstances outside their control had precluded its installation on one section for which it had been proposed.

Robinson made some interesting comments on the conduit system to the *Sun* in September 1902. He thought the slots were detrimental to carriages. 'As to motor cars, I don't know what the drivers will do when they come to negotiate the series of steel metals in South London.' He also thought the slots were liable to receive debris. The idea that the LCC was forced to adopt conduit because local authorities opposed the trolley system was unsupported by facts. Some twenty authorities in Middlesex and Surrey had agreed to overhead wires and the LUT had had trouble only with the LCC on that score. Such difficulties as the LUT had experienced with shallow mains in Twickenham and Teddington would be increased in proportion by using conduit.

Away in the far south-west of Surrey an electric light railway had been proposed to link Farnham and Haslemere via Headley and Hindhead. The LUT and the Imperial companies were interested. In November Farnham Council informed Robinson that several of its councillors supported the project. The idea came to nothing and a few years later the link was forged by an early motor bus service of the London & South Western. Even had the line been built, it is probably too fanciful to speculate on a grand Robinsonian trans-Surrey link with the LUT.

The *Chiswick Times* of August 28, 1902, gave an excellent pen-picture of Robinson as he was at that time. He was 'of middle height and would probably become portly if he took five minutes' rest a day, but as his days consist of about thirty hours, and there are some ten of them to the week, including Sunday, rest, even for the briefest interval, is unknown. Yet he looks well on a regime which would land most men in the grave or the lunatic asylum. Fresh complexioned, with clear eyes – they are of a dark grey – brisk and alert, he looks and is as fit as a professional acrobat. His manner is sharp and businesslike, but becomes enthusiastic when he talks of tramways, or takes you round the great, spotlessly clean powerhouse. . . .

'Mr Robinson was born in England on the banks of a Welsh river [*sic*]; his father was Scotch, his mother Irish, and he himself is a naturalized American citizen [*sic*], so, without hesitation, I give up the problem of his nationality. But he seems to me to have the phlegm of the English, the fire of the Welsh, the "canniness" of the Scot, the originality of the Hibernian, and the mechanical ingenuity of the American. . . . When he has any time to spare out of what I have called his thirty hours' day, he devotes it to the reading of exhaustive magazine articles, lecturing at professional institutions, or to any and every kind of outdoor sport or exercise – he is very fond of riding – except fishing.

'His son was apprenticed to him the day he was born. He is traffic superintendent of the LUET. Mystic letters, which once led a would-be classical scribe to remark, noticing them engraved on the power house: "Ah, I see you have a Latin motto up there. *Luet, luet,* let me see. Of, of course, Let there be light. Very appropriate – very appropriate indeed!"'

The needs of the Thames Valley extension called for the erection of a new depot. At first a site at Hampton Wick was considered but the choice fell on an 11-acre plot alongside Fulwell station on land leased from the Trustees of Sir Charles Freake. The site was excellent as there was ample room for expansion and, lying in the angle between the lines converging at Stanley Road junction, it allowed access from both Wellington and Stanley Roads.

A fine brick-built car shed, the largest on the system, was constructed in the centre of the site. The main part was double-ended and contained fifteen 400-foot tracks. There was also a small single-ended shed with three tracks on the south side. The total accommodation was for 100 cars. A substation adjoined.

In its October 15th issue the *Thames Valley Times* wrote that tradesmen in Broad Street and High Street Teddington would hail the approaching completion of tracklaying. The tearing-up of streets and resulting traffic congestion had been an annoyance for some weeks. The work was complicated by the fact that the roadway had to be lowered and new mains laid and tons of macadamite removed. For every mile of double line (as elsewhere) half a million jarrah blocks were laid.

But the same paper had earlier drawn attention to the great advantages already apparent in Twickenham, where, but for the coming of the trams, such improvements could not have come about for years, if ever.

The Board of Trade inspection of the Cross Deep–Stanley Road junction section took place on November 7, 1902. Several councillors turned up, to be told by Robinson that because of incidents during the inspection of the Cole's Bridge–Cross Deep section, he could permit only the chairman and officials to take part. After a wrangle, some councillors who had seated themselves in the inspection car got off. When the car reached the bridge carrying the LSWR over Heath Road, Councillors Webb and Scovell objected to the gradient of the dip, saying that it was too hard on horses.

The inspector politely pointed out that Parliament could allow gradients to be altered, but this one had not been. 'I will put the Royal Humane Society on to them [the LUT]' was Webb's parting shot. Not to be outdone, Scovell objected to the narrowness of Stanley Road and the siting of some of the standards.

The tour ended with an inspection of the transformer station at Fulwell. The line was passed and public service began on November 8th as far as the 'Nelson' at Stanley Road Junction.

About this time the *Middlesex County Times*, in its issue of December 20, 1902, ran a full page on Robinson. Work was his chief reward, it said. 'He would probably rather be given a free hand to organize London's passenger traffic than ten thousand a year and a seat in the House of Lords. The eight-hour day is not for him. Eighteen is nearer his mark, and into those eighteen hours he crowds more than most men would compass in eighty.'

The writer pointed out Robinson's extraordinary facility in getting work out of others. He could turn his hand to anything, from laying out the scheme to building the lines. 'He personally determines the car equipment, from the design of the apparatus to the buttons on the uniforms. He is also responsible for the building and installation of the power houses.

'His charming wife has helped him more than she knows, for the unfailing chivalry, the unfailing courtesy towards women which make them the readiest and most enthusiastic converts to his side of the gospel of progress have their roots in the reverence for her sex that his wife has inspired in him.'

A water drinker by choice, he had a temperament naturally exuberant and so vivacious that, to quote Oliver Wendell Holmes, 'if he took wine at all, it would be to keep him sober'.

His son discharged his duties with an aptitude and success only explained by the theory of mental heredity.

The article mentioned that tramway promotion under the 1870 Act had to be heralded by publication in the areas concerned twice in the same newspaper at intervals of a week. Under the Light Railway Act advertisements had to appear twice in different weeks but a week did not have to elapse before the appearance of the first and second notices. The writer instanced how Robinson used this difference to outwit rivals. So well did he and his staff prepare a particular scheme that when the rival schemes came before the Light Railway Commissioners no technical objections could be raised to what had been rushed through in four days and nights, while the rival scheme elaborated over months was rejected because of faulty estimates.

Chapter 8

Hampton Court and Uxbridge

Such progress had been achieved that a trial run was made over the Hampton Court loop on March 22nd. The *Chiswick Times* reported that the 'great new car' left Fulwell Depot at 9 p.m. 'A strikingly novel aspect it must have presented to the dwellers on the route as it passed through the comparatively open country between Fulwell and Teddington, throwing its halo of radiance across the meadows through the fast-falling rain.' Robinson junior was at the controls. With him on the platform were Robinson and the heads of the permanent way, overhead installation and rolling stock departments.

At Teddington, Hampton Wick and Hampton hundreds thronged the streets, for a water car had passed earlier and presaged the trial trip.

The report said that neither curves, nor gradients, nor the new Teddington railway bridge, nor the underline dip at Hampton Wick, nor the river bridge hump at Hampton Hill delayed progress. The 7-mile run took just over 45 minutes, including brief halts. The engineer (*sic*) pronounced it the most satisfactory trial trip he had ever accomplished. Robinson entertained the party with reminiscences of some that had not gone so smoothly!

The *Daily Telegraph* of April 1 waxed lyrical. 'It is a matter of calculable certainty that the townships through which the new line passes will reap, in higher ground and house rents, in increased rateable values and extended trade, more, vastly more, in the aggregate than the shareholders of the company will divide as profits.

'. . . There are those who think that the very sleepiness of Teddington and Hampton Wick is part of their charm. So it is; but nobody cares to eternalize narrow and nearly impassable lanes, and in broadening and straightening the roadway the company has done a permanent service, for which they have often been compelled to buy groups of cottage property and in some cases substantial buildings. . . . To be gifted by

Nature and not allowed to enjoy Nature's gifts is a crabbed contradiction from which the electric fairy is going, in part at least, to save us.'

The paper also said that the King had consented to a slice being taken off Bushy Park if the LUT put up new ornamental railings, and that Garrick Villa would be dedicated to the public enjoyment.

The Hampton Wick side of the loop officially opened on Thursday April 2, 1903, and the Hampton section two days later. Two hundred guests were invited to look over Chiswick power station and travel over the new route. They were received by Yerkes, as the new chairman, Samuel White and Robinson.

The party rode from Chiswick in four blue cars to Fulwell and then on to Garrick Villa where luncheon was served in a marquee in the grounds. There Yerkes formally declared the line open. He declared he had never seen a better 'outfit' than the LUT. The only fault was that the 'road' was too good commercially. The new Underground company hoped to do something to solve London's transit problem. The LUT was a sort of topping-off of their venture.

Robinson congratulated Yerkes on the happy circumstances in which he had made almost his first public appearance as chairman of the LUT. Yerkes, he added, had a long and wide experience of street railways but if pressed he would confess that for efficiency, substantial construction and luxurious equipments the LUT had few rivals and no superiors. Since electrification it had increased its passengers by almost 50,000,000, or a total of over 2,000,000 a mile annually – a figure probably unparalleled. Some 2,300 men were being employed on construction and operation and £240,000 paid in wages in two years.

The loop came into its own at Easter. On Easter Monday each car ran more than 130 miles. Cars ran from 7 a.m. to midnight at 35-second to 2½-minute intervals – on some sections there were up to 16 cars a mile. 250 permanent way men were deputed to help traffic regulation.

Robinson junior afterwards issued a general order: 'The Managing Director instructs me to express to the whole staff – engineering, operating and construction – his intense admiration for the splendid services rendered to the company during the Easter holidays, particularly Easter Monday, when all previous records were magnificently broken, a result which could only be achieved by the combined labour and devotion of all concerned.'

The Easter arrangements included circular tours out from and back to Hammersmith or Hampton Court, for 1s for the 24-mile trip. There was

Types 39 and 40. LCC cheap mid-day ordinary and workman tickets issued by LCC on route 89.

Top: Cars on inspection pit tracks in Fulwell depot. (*Courtesy R. B. Parr.*) *Centre:* Trams come to Teddington. Type W No. 249 inaugurates service in 1902. (*Courtesy Alan A. Jackson.*) *Bottom:* Hampton Wick about 1904. Type W (later U) No. 212 on Richmond Bridge–Hampton Court service.

Bottleneck on Hampton Hill, with type W (later U) No. 255.

Type W (later U) No. 266 passes the Swiss Cottage at Hampton during the Thames floods of January 1904.

Type W No. 173 at Hampton Court terminus. This car, still open-top, was remotored in 1927–8.

Hampton Court and Uxbridge

also a Richmond Bridge–Hampton Court–Richmond Bridge tour for 8d for 16 miles. A Kingston Bridge–Hampton Court shuttle service was put on, with 1d fare.

One press report relating to the Good Friday traffic ran: '. . . And of the one quarter million passengers it is safe to say that one half came from the Twopenny Tube, and a very large proportion from the East End of London. It is not very difficult to recognize the type which hails from Whitechapel way. The gorgeous hats with which the young ladies from the East throw down a challenge to their sisters in the West are well known, and if further proof of origin was needed yesterday, it could be found in the frequent interchange of head gear between young couples on the tops of trams, a form of pleasantry peculiar to Cockney lovers from the Mile End Road.'

Apparently the LUT could do no wrong. Over Whitsun the 300 cars by then in traffic carried no fewer than 800,000 passengers on all routes, compared with the 646,000 of Easter and the 405,000 of Whitsun 1902.

As the circular tours were so successful, the LUT made arrangements for private parties to hire tourist cars and the Immisch Electric Launch company's boats at Hampton. (This undertaking, established in 1887 and acquired by the BET in 1898, had the largest and best electric launch fleet on the Thames.)

In 1900 only three workmen's horse cars had been run, carrying 142,000 passengers. By 1903, with electric traction, there were 30 workmen's cars, carrying 2,562,000 at an average fare of under a farthing a mile.

Robinson was benevolent but still an autocrat and the paternalism at Chiswick was strict. Two men were discharged for attending a midnight meeting to form a branch of the Tramway Workers Association. The effort to form it failed. A drastic rule forbade men to attend such meetings on pain of dismissal. The men considered it an unwarranted interference with the liberty of the subject.

An open-air meeting was held in August under the auspices of the Gasworkers and General Labourers Union to air the alleged grievances of some LUT drivers and conductors. A *Chiswick Times* reporter could spot only one obvious LUT man by his uniform, but the meeting was not convened to address LUT staff *en masse*, Sunday being the day on which they could least be spared.

It had been impossible to form a tramwaymen's union. The men had to ask Robinson where they might go on off-duty evenings. If they attended a trade union meeting they were sacked, as in other industries.

All the same, the LUT had no trouble in recruiting staff. A notice 'Hands wanted' brought as many as were needed, even though there were complaints of 'pugilistic' tramway officials and of Robinson's wanting to make machines of his men.

Because of its heavy but sporadic traffic the LUT had to employ more men than it would otherwise need. Of a given payroll total only about 15 per cent were permanent, only another 10 per cent regularly employed. About 30 per cent were employed $2\frac{1}{2}$–3 days a week.

Sport was encouraged. In September 1903 the first annual LUT sports meeting took place on Pears Athletic ground at Isleworth. The events were arranged as a benefit for the families of three employees – a deceased inspector, and a sick inspector and gateman. The proceedings included swings, roundabouts, shows, dancing, fireworks and a concert. Robinson junior was one of the judges and Mrs Robinson presented the prizes.

In July the LUT was considering a claim by the owner of Cole Park Estate, Twickenham, for more than £1,200 for land taken, and incredibly, another £12,000 for consequential damage to the remainder. The LUT assessed the land at £760. A jury, to its credit, found for £759 for the land, with consequential damage nil.

In September four motormen appeared at West London Court to answer summonses taken out by Moffat Ford, managing director of the Motor Car Company, for exceeding 10 m.p.h. Ford said he had timed them at $15\frac{1}{2}$–17 m.p.h. with a stop-watch over 1 furlong. Cross-examined, he admitted he had been convicted – 'unjustly' – of driving his car too fast. The magistrate accepted a suggestion by LUT counsel for a Board of Trade test. When the case was re-heard in November, the drivers were each fined 40s plus 2s costs. Ford wrote to *The Times* maintaining that high speed with motor cars was safer than with trams. Robinson at once refuted him by pointing out that a tram was a known quantity keeping to a defined course.

In October a conference of some Middlesex authorities was convened by the Board of Trade to frame by-laws governing the LUT's operation. Some authorities wanted the trams to travel faster, some slower. A suggestion by Brentford that every tram should be disinfected once a month was ruled out as *ultra vires*. Robinson offered £500 to anybody who would invent a reliable speed indictaor.

The LUT Act of 1903 sanctioned no extensions, the Bill having been promoted to give powers to make widenings and other improvements in connection with authorized lines to a total cost of £80,000. By contrast,

Hampton Court and Uxbridge

the Bill for the 1904 Session included ambitious projects: extensions from Baber Bridge to Bedfont and Staines (Market Square) and (a revival of the 1902 project) from Cranford to Colnbrook, Slough and Maidenhead Bridge (by Skindle's Hotel), and a tube tunnel, large enough for double-deck trams, from the Surrey side of Hammersmith Bridge to a sub-surface terminus adjoining the District Railway platforms at Hammersmith station.

Also included were a link between Brentford and Ealing via Ealing Road (reviving an earlier project) and a 3-furlong line over the new Kew Bridge, which had been opened on May 20, 1903. The total proposed was 21½ route miles, on which the LUT was prepared to spend £217,932 in road and bridge widening.

The Maidenhead proposal, like that of 1902, included three-quarter-mile of new road at Colnbrook, with a bridge to avoid the level crossing. For the 1 mile 7 furlongs in Slough the LUT was ready to spend £31,364 on widening to 50 feet between fences. But Slough Council not only demanded hardwood paving from kerb to kerb, at an extra cost of £19,969, but also conduit operation within its area so that the beauties of the high road through the town should not be impaired by wires, a demand which would have cost another £45,769.

Elsewhere in Buckinghamshire the company was prepared to spend £29,688 on road works, but the county council's insistence on a 40-foot carriageway throughout, with wood paving, would have increased this sum by about £30,000.

But even these demands paled before the rapacity of Brentford. In return for its assent to the Maidenhead extension, which admittedly, like that to Staines, would have increased tram traffic through the town, Brentford council required the LUT to widen the High Street throughout to 80 feet! It also demanded £1,000 a year rental and the construction of a lavatory and subway. By its 1898 Act the LUT had committed itself to £40,000 worth of widenings and other works in the town. But the wholesale widening of the High Street would have cost it no less than half a million!

Because of such extortionate demands the Brentford–Ealing proposal was doomed too. The LUT was ready to widen Ealing Road from 21 feet to an over-all width of 50 feet. Widenings in Ealing would have cost £13,536, but Ealing demanded £7,000 more of extra works. The length of Ealing Road in Brentford was less than a mile. The LUT was ready to spend £36,659 on widening it from 21 to 32½ feet. Brentford council made demands which would have cost another £520,000!

Little wonder that faced with such an attitude the LUT was forced to drop from the Bill the Cranford–Maidenhead, Ealing–Brentford and Kew Bridge proposals.

But the LUT was not yet finished with the almost vindictive opposition of local authority. The Examiner on Standing Orders rejected the Hammersmith tram tube project on the allegation by the LCC that it should have been proposed as a railway not as a tramway – the cost of the twin tunnels was about £500,000 and that of the short surface linking tramway, £1,000. Robinson rightly commented that such an allegation came with ill grace from County Hall in view of its own tramway subway proposals.

All that was salvaged from the Parliamentary wreck was the Staines extension, estimated to cost £109,605 and not seriously opposed. This, with an extension of time for the authorized Surrey lines and authority for widenings to remove a number of single-track sections, was sanctioned by Act of August 15, 1904.

Robinson suggested it was high time for a thorough revision of the powers of veto of local bodies. But his chief scorn was reserved for the LCC whose action had set back at least two years the execution of a scheme 'admirably conceived by Yerkes', to provide easy tram-train interchange at platform level at Hammersmith and so wasted the sums spent on surveys and the preparation of Parliamentary plans.

But traffic remained buoyant. At the annual general meeting in February 1904 Samuel White said that the Askew Road line would have a far-reaching effect on the LUT, particularly as it was now associated with the District Railway. He considered the outlook exceedingly satisfactory. The latest gross receipts were £280,242 and the working and general expenses and renewals totalled £176,392. Compared with the 36,000,000 of the preceding year they had carried 45,293,473 passengers. When the weather was good for pleasure traffic, receipts invariably exceeded the estimates. A final dividend of 8 per cent was declared on the ordinary shares for the half-year to December 31, 1903.

Work had begun on July 27, 1903, on the Southall–Uxbridge line. Robinson and members of the engineering staff made a trial run between Southall and Uxbridge and over the Askew Road line on March 28th. The car left Chiswick at 8 p.m., reached Uxbridge at 9.45 and returned at 10. The Board of Trade inspection took place on both sections on Tuesday May 31st and public service began the following day.

Hampton Court and Uxbridge

The LUT directors entertained Colonel Yorke and officials and members of the councils served by the Uxbridge extension to a sumptuous luncheon at the Chequers, Uxbridge, where the Uxbridge and Hillingdon Prize Band was in attendance.

The publicity department was kept busy. Two handsome illustrated books were produced for the LUT by W. T. Pike & Co., of Brighton.

One, 'To Uxbridge from the City by Train, Tube and Car via Ealing or Harrow' begins with a potted history of the areas traversed not only by the LUT but also by the railways under construction by the Great Western, via Greenford, the District (Ealing–South Harrow extension) and Metropolitan (Harrow–Uxbridge). It goes on to describe the new developments at Ealing, after mentioning the 'superior' class of houses erected on the Elms Estate at Ealing Common: 'Until recent years few new houses were built in Ealing of a less rental than £70 or £80 a year, and a considerable portion of the area of the borough is occupied almost exclusively by houses with rents ranging from the figure mentioned to £200 a year and more. But in South and West Ealing cheaper houses from £70 to pretty semi-detached villas at £30 have been and are being built in response to a public demand that shows little sign of abating. While the electric trams have not driven the richer class of inhabitants from Ealing, they have made the borough a pleasant place of residence for those less well endowed with the world's goods.'

Building is said to be in full swing at Southall and the electric car has also enlivened Hillingdon.

The advertisements of public houses such as the 'Waggon & Horses' at Hayes and the 'Ye Olde Treaty House' at Uxbridge stress their nearness to the tramway.

Of Uxbridge the booklet says '. . . the electric car has transformed what hitherto was the quiet extremity of a busy little place into a region of mighty throng – to the great pecuniary advantage of those shopkeepers whose business has any connection with refreshments. . . . It is a wonderful fivepennyworth of locomotion, this ride from London to Uxbridge. No omnibus ride can approach it: Charing Cross to Finchley, Barnsbury to Tulse Hill, Highbury to Putney are all long rides, about eight to nine miles, four or five less than the run to Uxbridge, and the fare for those and similar trips is sixpence. . . . These cheap fares mean that many persons see the country who, otherwise, would not – who would not, or, at least, not so soon, see "a rabbit with the skin on", as a child exclaimed after enquiring what certain moving little objects were in a field at Hayes.

'Not that everyone is altogether pleased with this rush to Uxbridge. The place has been likened to Pandemonium and calculations are made that persons are shot out here at the rate of a hundred every three minutes. Twenty-five thousand people are turned loose here on a Sabbath day (the Hampton Court tramway takings being beaten by £70 in one day). . . .'

'What they [the LUT] have done, extensive as is the work, it is only a portion of what is intended if absolutely impossible conditions are not imposed. . . . In constructing the extension to Uxbridge, the London United Tramway Company has had no such battle to fight as fell to its lot when it set about making the extension from Acton through Ealing. Uxbridge wanted the tram from the first and neither Hayes nor Hillingdon offered any organized opposition.'

A short biography of Robinson was followed by a note on Robinson junior: 'Mr Clifton Robinson, junior, born at Bristol 1880, is Superintendent of the Lines, and probably the youngest man occupying such a responsible post in Great Britain. He has been Superintendent since September 1899. He is intensely practical, a perfect master of the details of his business, and has the distinction of having driven the first electric car in London over the Kew and Acton lines, in February 1901. For the past seven or eight years he has been an ardent motorist, and his 60 h.p. Mercedes racer is a familiar object on his daily tours of inspection over the extensive routes of the London United Tramways.'

On the back cover of the Uxbridge book was a rather crude representation of a tram and some of the places served, above an advertisement for the new services, with such words as 'Delightful Excursions' and 'Motors for the Million'.

The companion booklet is entitled 'By Tube and Electric Car To Hampton Court'. It is equally lavishly got up, with many illustrations and is valuable for general as well as particular information. It confirms that at that time all white cars were allocated to the Ealing and Uxbridge line, red cars to Kew and Hounslow and blue cars to Kew, Hampton Court loop and Richmond Bridge services. All cars from Shepherds Bush or Hammersmith to Hampton Court travel via Hampton. The Teddington and Hampton Wick side of the loop is served from Richmond Bridge, though the arrangement is varied on summer Sundays and Bank Holidays. At night the cars are distinguished by lights – those going via Teddington carry a mauve light.

Hampton Court and Uxbridge

Of Garrick Villa the booklet says that the grounds extend far to the rear. 'In a distant field we have seen 1,000 wives and children of LUT men taking tea in a tent, with Mr and Mrs Clifton Robinson competing to secure the fattest of babies for brief periods of nursing.'

Of the negotiations to get the tramway through Brentford the booklet says: 'Yerkes in Chicago, Whitney in New York or Johnson in Cleveland probably never accomplished a greater feat of negotiation than that which effected this result.' But some fresh solution would have to be found, either a widening or the construction of a new street for eastbound traffic.

A much-needed track doubling was promised when the widening of High Street Hampton was completed.

Some interesting future trans-Thames possibilities were suggested. The proposed tram tunnel under the river at Hammersmith (as yet still a possibility) would give the LUT its longest run without a stop and the 'ride outside' beneath the Thames would give Londoners a new sensation. It was possible trams would cross the river at Richmond and Kingston on special bridges, though the widening of the existing Kingston bridge was more likely. There might be a tramway bridge close to Corporation Island at Richmond.

The Uxbridge line would have opened sooner but for some delay in completing the depot and substation at Hillingdon Heath. The substation resembled those at Hanwell and Hounslow. It had two 250 kW. rotary converters and seven 100 kW. transformers. If necessary, it could be fed from Hounslow. The depot, with true Robinsonian lavishness, measured 150 feet by 100 feet and like those at Acton, Hanwell and Hounslow, could house fifty cars, in this case twice the number expected to serve the route. And the LUT had no plans for extension beyond Uxbridge or branches from the line! It is a pity that some of the money spent on the depot could not have gone into laying double track throughout, instead of single line with loops for much of the way beyond Southall.

The opening of the Uxbridge line, and that along Paddenswick and Askew Roads, required an extra 200 men on the regular traffic staff, making a total of about 750 for the northern part of the system and about 1,700 for the whole system so far opened.

The new line was the scene of a bad accident in December 1904 when one evening a car coming from Uxbridge collided with a trap on Hillingdon Hill. Three cars were brought to a stop on the single line. At the top of the hill, some way off, a fourth car from London ran into

London United Tramways

another and both careered downhill to collide with the three cars at the bottom. One car was swung across the track and others were badly damaged. A driver and some passengers were injured and the line was blocked until next morning.

In spite of poor weather Whitsun traffic was good. On Whit Monday all cars were out and some 350,000 passengers were carried, of a total of 675,000 for the whole holiday.

In April the company won an appeal for assessment by Brentford, the net amount being fixed at £11,225. The expectation of life of cables was given as 32 years and that of the overhead as 23 years. The cars were expected to last 14 years – they would be made to last a great deal longer!

In declaring an interim ordinary dividend of 6 per cent for the first half of 1904, the directors said they considered it a good policy to provide at once for the development of traffic rather than delay improvements and widenings which would become much costlier with rapid increase of property values along the routes.

In his evidence before the Royal Commission on London Traffic in June 1904, Robinson said that the LUT had 36 miles in operation, 38 more authorized and another 5 just passed by a Commons Committee. In 1900 they owned 50 horse cars and received £59,000 in receipts, compared with the £278,000 in 1903. In 1899 the population of the areas now served was 424,100, compared with the present 533,709. The following were some of the places quoted:

	1899	1904
Southall	7,627	13,200
Ealing	23,978	33,040
Twickenham	16,206	20,991
Teddington	10,025	14,029
Heston/Isleworth	26,271	30,836

Robinson spoke of the easier procedure in the United States for gaining permission for tramways. Clause 43 of the Tramways Act was most pernicious. He believed that the Act had been passed primarily to enable the Board of Trade to grant Provisional Orders. It had been more by accident than design that local authorities had begun to build and work tramways.

If the LUT had been less squeezed by local authorities the fares could have been set lower still. It cost between £300 and £400 a year to maintain a mile of tramway. Most of the sum was spent simply to restore the

Hampton Court and Uxbridge

wood paving damaged by the rutting action of ordinary traffic alongside the rails and had nothing to do with the running of the cars.

Parliament should be recommended to abolish the local authorities' right of veto, conferred at a time when there was no idea of electric traction or of local authorities becoming rival promoters and operators. Since 1898 the company had had to abandon 59 miles 5.9 chains of proposed lines because of the attitude of local authorities.

The London County Council at first would not entertain electric traction or overhead wires. Yet as soon as the LUT got its first line going the council rushed to the other extreme and wanted tramways everywhere. He himself wanted to run the LUT right into London but would be content to form junctions with the LCC and grant reciprocal running powers.

Tramways were obstructive only where stub termini existed. They had obviated this at Hammersmith by building a loop.

The extension of tramways into Central London would remove much other traffic from the streets. Why London 'where tramways are more necessary than anywhere else' was without them he did not know. Given roads sufficiently wide, a tramway would be of the greatest benefit to ordinary traffic. A width of 33 feet from kerb to kerb was the desirable limit for tramways to be laid down.

There were uses for the conduit system but they were circumscribed. It was costly to install and if anything went wrong the street had to be torn up.

The only LUT line so far built under a Light Railway Order, under which local authorities had no power of veto, was between Southall and Uxbridge, but their proposals included lines to Kempton Park and Hanworth, which had not yet been confirmed.

In May 1905 Charles James Cater Scott, chairman of the London & India Docks Company, was elected a director of the LUT and appointed chairman in succession to Yerkes, who wished to give more time to the District Railway. Yerkes died on December 29, 1905.

One of Cater Scott's first acts as chairman was to preside at a meeting of the directors on July 3rd when it was resolved 'that the Board desires to congratulate Sir Clifton Robinson upon the honour conferred on him by His Majesty the King, in recognition of his lifelong services in the work of tramway development, and the directors are glad to place on record their hearty appreciation of the well-merited distinction bestowed upon the company's Managing Director and Engineer'.

The knighthood came at the peak of his business and social career.

The year before he had become a Justice of the Peace for the Willesden Division of Middlesex. He was a member of St Stephens Club and an Associate Member of the Institute of Civil Engineers and the Institute of Electrical Engineers – he became a Member of the IEE in 1908. In November 1906 he was admitted as a Freeman of the City of London by redemption (purchase) on the Company of Makers of Playing Cards. He was also elected a Fellow of the Royal Colonial Institution.

As well as Garrick Villa he now had a town house (Keith House) in Porchester Gate. The location is significant. It was close to the terminus of Train's short-lived Bayswater Road of 1861.

His recreations appear as travel, music, motoring, outdoor sports, horses and dogs. His close friends included W. J. Bull, Sir Thomas Lipton and William Whiteley. Visitors to Garrick Villa were drawn from a large circle and included Bernard Shaw and Louis Wain, the cat artist.

He had a fund of stories and a ready wit. When he and his wife inaugurated the miniature rifle range for LUT men at Chiswick both scored a bullseye with their first shot. The result was adjudged a tie. 'One more to the many that unite us', commented Robinson.

Before he took up motoring Robinson liked to ride in a smart pair-horse turnout driven by Billy, a ruby-faced coachman, who was trained to be a chauffeur. There were two other pairs of carriage horses.

Midnight on October 31st found Chiswick depot once more in festive array. The staff had come to do honour to their knighted chief. The depot was floored over, hung with tapestries and decorated. Robinson presided, with Lady Robinson at his side. Others there included Cater Scott, Sir James Szlumper (as Mayor of Richmond), the Mayor and Mayoress of Kingston, Clifton Robinson Junior and W. J. Bull.

More than 1,300 sat down to supper. Divisional Inspector Newby read an address of loyalty, esteem and respect of the staff to Sir Clifton, Lady Robinson and Clifton Robinson junior. At Sir Clifton's hands they had received the utmost consideration and justice and unceasing kindness. They felt pride at his distinction, which added a dignity to electric traction enterprise. Their feelings were more deeply expressed in their personal affection for 'The Chief'. Lady Robinson accepted a silver writing set from the staff.

Cater Scott said that Robinson's evidence before the Royal Commission on London Traffic had shown his masterly grasp of the subject. The LUT was trying to get people out of the crowded whirlpool of London to places where they could breathe the fresh air of heaven.

Hampton Court and Uxbridge

Bull declared that Robinson had won his knighthood as truly and bravely as any man who had been knighted on the field of Agincourt or Crecy. He himself had been a diligent passenger on their trams for a number of years and had always received the greatest courtesy and attention. He had a cordial admiration and affection for Robinson and had received many kindnesses from him. He had known hundreds of instances of Robinson's kindness to his fellow men.

Robinson told the gathering that in defiance of all obstacles they had carried the record number of nearly 400,000 passengers in one day. It showed the excellence of their installation and the energy and efficiency of their men that the population of a huge city could be carried conveniently, cheaply and comfortably out to green fields and the banks of the Thames without injuring a soul or receiving a word of complaint. His own distinction he took to be a much broader guerdon than for his attributes.

He hoped the new chairman would come among them on many future occasions. He would promise him a delightful time on the football ground at Brentford, when the men he saw before him renewed on the green field their manly sports and pastimes. When the LUT tug-of-war team had gone to York to meet the pick of the country's tramwaymen they had pulled them all over the field and each man had come back with a silver trophy.

LUT men had made the LUT one of the most appreciated tramways in the country. From the 56,000,000 passengers entrusted to their care in a year, he had received hardly any well-sustained or well-established report to discharge one man for misconduct. No man had ever been reprimanded for 'unsteadiness' or other idiosyncrasies which in the past had prevailed among tramway workers. They were steadiness itself. The doctor said they were in good health. They had eager faces and good hearts and he admired every one of them.

Receipts for 1905 were £301,350, a rise of £5,114. Expenses were £874 down, at £175,313. An ordinary dividend of 4 per cent was declared for the first half of the year and of 2 per cent for the second half.

The LUT had almost shot its bolt with new lines. Its Act of June 30, 1905, merely granted extensions of time for certain sections authorized in 1901 and 1902. That of August 8, 1906, was for widenings, acquisition of land and other purposes. The 1901 territorial agreement with Middlesex County Council was superseded by a new agreement which fixed the boundary at the Shepherds Bush–Uxbridge line.

Chapter 9

Into Surrey

The estimated cost of the new Surrey lines was at least £700,000. J. G. White & Company gained a £165,000 contract to build and equip $12\frac{7}{8}$ miles, including the Brentford–Hanwell section. John Mowlem & Co. Ltd gained the Wimbledon–Tooting contract.

Construction of the Kingston area lines began on April 3, 1905, in Kingston Road, Malden. Robinson was there, accompanied by Alderman H. C. Minnitt, Mayor of Kingston, and Mr A. Streeter of the Maldens & Coombe council. After Robinson had briefly explained the scheme, Alderman Minnitt wielded the first pickaxe as a signal for 400 men to begin digging up the road.

The construction involved some heavy works. There were many sharp curves; one of more than a right angle was the turnout from Claremont Road to St Mark's Hill in Surbiton, at the foot of a steep gradient. The road under the LSWR bridge in Richmond Road, at Kingston Station, had to be lowered by 2 feet 6 inches, and retaining walls built. Under the narrow skew arch in Kingston Road, Malden, the road had to be lowered 4 feet 6 inches, four 30-inch water mains lowered and a new sewer laid.

At the junction of Kingston and Malden Roads, Malden, Derby Lodge and its garden had to be removed to give a straighter run.

On the Malden–Wimbeldon–Tooting extension a new bridge was built to span Beverley Brook in Burlington Road. West Barnes Lane was widened from 25 to 45 feet, and an avenue of 100 elms felled. The bridge carrying the LSWR main line over the lane was rebuilt at a cost of £7,000.

Altogether on the Kingston area extensions there were 31 separate widenings, of which 29 were carried out by J. G. White & Company.

Track on these and other new lines consisted in the main of No. 3 British Standard 100 pounds a yard rails, $6\frac{1}{2}$ inches deep and 7 inches wide at the base. The joints were formed by H section anchors, 30 inches in length, 5 inches wide and 8 inches deep, weighing 30 pounds a foot, and

Into Surrey

a steel plate 30 inches in length, 12 inches wide and half an inch thick, fastened under the rails by hook bolts, the plate being between rail and anchor.

The attachments were made by the hooks of the bolts fitting into slots in the upper flange of the anchor, the bolts passing up through holes in the plate and bottom flange of the rail, and secured by nuts, screwed down on to a bevel washer.

At 9-foot intervals a sleeper consisting of a 6-foot length of old girder rail with a depth of 6 inches and a flange of similar size was embedded in the concrete, with flange uppermost. Two seven-eighths-inch bolts with bevelled washers secured it to each rail.

Manganese cast steel was supplied for points and crossings by Edgar Allen & Co. All points were laid 9-feet 6 inches long on tread and were drained. Joints were bonded with two Crown copper bands, 32 inches long. The track was cross-bonded every 40 yards of double track and the two sets of rails cross-bonded every 80 yards. After platelaying and bonding were completed the track was concreted in and then paved between the metals and for 14 inches outside each outer rail with jarrah blocks, 9 inches in length, 3 inches wide and $4\frac{1}{2}$ inches deep. Outside the wood paving was laid an edge of best Aberdeen granite setts.

Great care was taken with drainage and in places the concrete bed, always at least 6 inches deep, was taken down to 15 inches. The bed was finished off with a 1-inch layer of fine concrete. On top of this came a 1-inch floating coat of cement and sand to form a bed for the blocks. Double track cost about £13,000 a mile.

Span wire construction was used throughout, with the wire at 23 feet above rail level, save under bridges.

Standards were 33 feet tall and ornamented. The trolley wire was 2/0 B and S grooved copper with mechanical ears as supports.

W. T. Henleys Telegraph Works Co. received one of the biggest contracts so far placed for tramway cabling. It was for some 30 miles of 3-core, 3-phase, 11,000-volt cable and about 45 miles of low-tension feeder, distributor, telephone and other cabling. Two h.t. cables were laid from the Lots Road, Chelsea, power station of the Underground Group, over Wandsworth Bridge, through Wandsworth and Wimbledon Park to the tram route in Wimbledon and then alongside the route to Kingston.

The Kingston/Surbiton area was fed from a substation in London Road, Kingston, which housed seven 200 kW. oil-insulated, self-cooling transformers and two 500 kW. rotary converters. A second substation in Salisbury Road, Wimbledon, housed three 500 kW. rotaries and ten oil-insulated tranformers. All equipment was British Westinghouse.

Chiswick power station supplied current at 500 volts direct to the original electrified routes. The substations at Hanwell, Hillingdon, Hounslow and Fulwell transformed 5,000-volt current from Chiswick to 500 volts for the Uxbridge, Hounslow and Thames Valley extensions.

At 11 p.m. on Sunday, February 11, 1906, two of forty handsome new covered-top cars with end balconies left Fulwell Depot. The first, No. 309, carried Robinson and other officials, the other a gang of workmen. At midnight Robinson took the controls of No. 309 at the new turnout in Hampton Wick, to become the first man to drive an electric tram over a Thames bridge.

Five hundred spectators braved the winter night to cheer him. Their hardiness was rewarded, for No. 309, the finest public service vehicle yet seen on a Surrey street, must have been a brave sight as, lights ablaze, it moved over the still narrow Kingston Bridge and on into the town. Many people tried vainly to climb aboard as the car went down Clarence Street. On the way back, three policemen and a well-known lady resident were among those who managed to clamber on.

Surrey had been stormed at last. Ten months of street dislocation – five longer than the company had expected – were over, and the *Surrey Comet* could cease its weekly editorials on the vexed subject. Even so, Kingston Corporation needed a few more reassurances, and there was more discussion with Robinson and Surbiton Council before it was agreed to start public service on most of the new routes on Thursday, March 1st.

On February 21st Lewis Bruce, a coloured man who was Robinson's 'personal' driver – though he also drove service cars – successfully piloted the Board of Trade inspection car, No. 321, over the new system. Aboard were Robinson senior and junior, W. G. Verdon Smith (LUT Secretary), LUT engineers and local officials.

Colonel Yorke was pleased to learn that the LUT would lessen the danger to pedestrians on Kingston Bridge by agreeing to carry them over the bridge at no extra charge.

The car went first to Kingston Hill, then returned to Eden Street for the run to Surbiton, where photographs were taken, and Tolworth. There was a slight hitch on the way back from Tolworth to the Dittons when an oil switch at the substation failed and halted the car.

Into Surrey

Back at Hampton Wick Colonel Yorke congratulated Robinson on a first-rate road.

Although March 1st dawned dull, Kingstonians were not dismayed. Many shopkeepers had shut for the day and had added their own decorations to the garlands that the LUT had raised on Kingston Bridge and its approaches. Enthusiasm mounted as, under Robinson's supervision, the Mayor drove No. 320 – first of three bedecked cars crowded with LUT officials and distinguished guests – set out at 11.30 over the bridge, through Clarence Street, along London Road and up Kingston Hill to the end of the line by the 'George & Dragon'.

The Mayor handed over to Bruce for the tour of the remaining routes. As the car began to descend to Norbiton, two of Hodgson's Brewery drays were lumbering up the slope. The first dray passed but the second horse swerved, so that part of its dray caught the front of the tram. Robinson, jammed between the tram and the dray, fell back into the road, spraining his ankle and sustaining bruises. Although in pain, he insisted on completing the tour. The dray was damaged and the brass rail, a window and some woodwork of the tram suffered.

After the tour, 200 people sat down to a fine *déjeuner* in Nuthall's Restaurant, Kingston. The LUT was, as usual, adept at staging the inaugural junket. Silk ribbons on the tables represented the tracks. Slim vases with tulips simulated the trolley poles and lengths of smilax, the wires. Each guest received a buttonhole.

Acerbities were past. The Mayor complimented the company handsomely on its enterprise, determination and dogged perseverance. 'May they prosper greatly, may their dividends go up and may they always be able to keep their fares down, may their cars never lack passengers and may their passengers never have to look for cars.' His Worship understood that Sir Clifton intended to call his new cars 'whizzers' because of their speed. Only on that condition would they condescend to notice the existence of motor buses by the name of 'buzzers'. Sir Clifton had shown he was the wizard of the LUT. [Those 'buzzers' would show the LUT their sting in due time – already the London & Suburban Omnibus Company, one of whose routes linked Kingston and Richmond, had bought new Crossley-Leyland motor buses.]

Cater Scott expressed the hope that Kingstonians would soon forget the inconvenience they had endured. Sir James Szlumper, as Mayor of Richmond, rose to say how much he regretted his town's attitude to the LUT. [His own had changed too!] He was sure the trams would bring enormous numbers of people to shop in Kingston. Then the Chairman of Esher & Dittons Council twitted Robinson about 'Winters Bridge' being shown on the cars as the name of the terminus in Long Ditton and hoped it would soon be righted to Windows Bridge.

Robinson, in spite of discomfort, was in his usual form. Szlumper, an engineer, had praised the track, he observed. Well, there had been no jolt from end to end, save for the one he himself had got! In the streets and roads of Kingston and Surbiton, they had laid 4,500 tons of rails, sleepers and fastenings, 6,290 tons of cement, 50,700 tons of ballast and 1,000 tons of granite setts. The company had already carried 201,720,930 passengers electrically – almost 40,000 a day – and gained a total revenue of £1,200,479, all from an average fare of less than $1\frac{1}{2}$d.

Characteristically, there was praise for the staff – smart, attentive and rarely complained of – and a word for Edgar Speyer for finding the money for London United expansion. 'When any little tangle occurs,' Robinson wound up, 'ring me up, and I shall respond cheerfully.'

Public service began the same day to Long Ditton, Tolworth and Kingston Hill. The other routes, along Richmond Road to Ham Boundary, with the branch up Kings Road to Richmond Park Gates, and the first part of the Kingston–Wimbledon–Tooting route, from Norbiton Church along Cambridge and Kingston Roads to Malden Fountain, followed on May 26th.

The *Surrey Comet* of April 7, 1906, wrote: 'Time . . . now sees the omnibus ruthlessly swept away before the irresistible onrush of the field of electric traction. How many of those who looked on the growth of the omnibus service in the district from the time twenty years ago, when the first one-horsed omnibus ran from Kingston to Surbiton, expected to see them rendered useless by the coming of the swift-running electric cars ?' There was a reference to an auction of the Kingston–Esher horses and buses – in spite of an advertisement reading 'admirably adapted for motor power', three of the four buses, the newspaper reported, had gone for absurdly low figures.

The price for the latest advance was stiff. For the almost 12 miles of line the LUT had to find £66,000 for the local authorities for street improvements and undertake to contribute £10,000 each to the cost of rebuilding Kingston and Hampton Court bridges if the bridge authorities considered that necessary for tramway operation.

Some of Malden's £1,500 exaction was to meet the cost of an 'improvement', not necessarily on the tram route, to be carried out before the

Left: Sir James Clifton Robinson. (*Courtesy Florence Comerton*). *Centre:* Entertainment by an LUT 'serenader' in Garrick Villa grounds, Hampton. (*Courtesy Florence Comerton*). *Bottom:* An LUT band plays in the grounds of Garrick Villa. (*Courtesy Florence Comerton*)

Top: Holiday crowds by the river at Hampton, probably on the occasion of an LUT staff outing to Garrick Villa. (*Courtesy Florence Comerton*)
Centre: Guests of Sir Clifton Robinson take the air at Garrick Villa. (*Courtesy Florence Comerton*)
Bottom: Sir Clifton and Lady Robinson at Garrick Villa. (*Courtesy Florence Comerton*)

Top: Type W No. 196 on Hampton Court–Hammersmith service about to pass a car on Richmond Bridge–Kingston service, possibly at Twickenham Green. The presence of the buses, one a General 'B' type, suggests clearance tests. *Centre:* In Broad Street, Teddington. *Bottom:* Widening in progress by Peg Woffington's cottage, Teddington.

Top: Disruption in Eden Street, Kingston, for LUT tracklaying in 1905.
Centre: Triangular junction at Kingston Bridge approach, Hampton Wick.
Bottom: Hampton Court terminus, with type W No. 296 in foreground.

company began any work! In addition, Kingston wrested an annual 'wayleave' payment of £225 for five years, £450 for another ten and £700 for a further ten years.

Esher & Dittons demanded £150 similarly and the Moleseys – where no LUT car was to penetrate – £100. Surbiton's price was £300 for five years and £500 a year thereafter, and the Surrey County Council's was £487, representing £100 per mile of main road traversed. Such demands when capitalized totalled more than £68,000.

Small wonder that only a few more miles would have to round off Robinson's empire and that no trams would ever carry crowds to the Derby or Sandown Park, as Robinson had dreamed.

Perhaps Kingstonians soon had cause to ring up Robinson, for after a year or so they lost their new trams. Heavy on power for the Surrey routes, fed from Lots Road, Chelsea, they were exchanged for an equivalent number of open-toppers from the Uxbridge and Hounslow routes.

With their low fares and frequent service, the LUT trams did much to confirm Kingston as the shopping centre for a wide area. Clarence Street and Richmond Road began to take on a more modern aspect. Soon the cottage on the corner of Richmond Road and London Road gave place to a cinema, and across the road the Empire Music Hall arose to provide added patronage for the trams.

The Boston Road line also opened on May 26, 1906, though with less éclat. With the final sanction of the Acton–Hanwell line, the need for its early construction had lessened, as it no longer formed part of a trunk route.

The extension to Hook and the Dittons–Thames Ditton–East Molesey–Hampton Court loop were never built.

In February 1906 Robinson wrote to the Board of Trade about complaints of noisy running made by the borough surveyor of Ealing. Forty-six cars had been withdrawn for alteration of brake rigging. The sixteen so far altered, said Robinson, had shown the same quiet running as the altered car demonstrated to the Board of Trade inspector. 'I take the strongest exception to the statement that the noise is now "as bad as ever" or that there have been any grounds for reasonable complaint since the date of the inspection'. A superficial inspection would show the council the improvements in Ealing and elsewhere where corrugations on the rail tread were being removed by grinding.

They were introducing a new type of trackwork using anchors and anchor joints, which was an improvement on the old. The main object was to prevent vertical movement of the rail under traffic, which was detrimental to the roadbed. A new method of track repair which they had been using for more than a year was also proving an advantage. They were trying to devise a system of thermit-welding the joints to strengthen the rails still more where it was most essential.

Chapter 10

Rounding Off

The Tatler for December 5, 1906, contained a fascinating contribution by Robinson entitled 'Tattle about Tramways'. It is particularly valuable as showing that, in spite of his predilection for tramways, he kept an open mind.

He was satisfied that London had to look to electric traction for a true solution of the traffic problem. He had no prejudice against the motorbus *per se*. Where there were no tramways or where for the moment they were deemed impracticable there was ample 'verge and room' for the bus. But he saw no reason to depart from the opinion he had expressed in *The Times* in April 1905 that 'properly developed it might fulfill a useful mission as a feeder to electric tramways but it was too much to expect that such a novel and complicated box of necessarily tabloid machinery could at one bound encounter and overcome the trying conditions of actual road surface work or in such a way hope to hold its own in direct competition with established electric tramways. . . . That the doom of the motorbus as at present designed has been pronounced is generally conceded'. Time and ingenuity would show whether a really serviceable public vehicle would be evolved from the wreck.

'When it can be shown that any system other than electric traction can maintain its claim as a successful competitor, then I shall realise that the supersession of electric tramways has come within the sphere of practical politics.' But such a system had to be capable of universal application.

These remarks must be set against his comments of a year or so before when he told a reporter: 'As against an electric car, it [the motorbus] has about the relative value of a perambulator.' It cost 6d a mile to operate a tram, but 1s to operate a motorbus. A tram motor lasted fourteen years, a bus engine two. A tram could seat 70, a bus 32. 'Our extra earning capacity will more than cover the initial cost of the track and its maintenance.'

In February 1907 *The Railway Magazine* accorded Robinson one of its 'Illustrated Interviews', normally granted only to railway chiefs. Robinson, thought the interviewer, might well be styled the Apostle of the Light Railway. Of Robinson's tramway initiation at Birkenhead: 'With a keen adaptability of humour to business conversation, which is one of the charms of Sir Clifton's personality, he had told how he "pursued that car, got on to it, held on to it, and from that day to this I have never let go of tramways".

The account, which also usefully summarized the development of the LUT, related how one day a 'father of the vestry' asked Robinson if he would provide waiting rooms as passengers would not care to wait twenty minutes for a tram. Robinson replied that they would be unnecessary as the service would be continuous. The vestryman half believed Robinson was making fun of him.

The Malden–Raynes Park section was opened on Saturday April 27, 1907, after an initial setback when Colonel Yorke would not at first approve the wood paving in Burlington Road. Many Wimbledonians, who must have wondered if trams would ever reach them, 'went for an airing' – as the local paper put it – from Raynes Park to Malden or Hampton Court. Robinson had hoped to open right through to Wimbledon. The track had been laid through Wimbledon for almost a year but the council pressed a clause in the Act stating that the whole line to Tooting had to be ready before they would consent. Robinson answered with a completion date well within the statutory time and offered a bond of £20,000, to no avail.

Robinson was galled by the delay. Among other reasons he was anxious to tap the District Railway at Wimbledon without delay. The council relented sufficiently to allow operation from Raynes Park along Worple Road to the Hill Road end of Worple Road from Thursday May 2nd, subject to the Board of Trade passing the line and the LUT undertaking to complete the rest of the associated widening and other works as soon as possible.

The *Wimbledon Boro' News* of May 4th said: 'In allowing the trams to run at once on the Worple Road line only, and deferring the using of the section to the Grove [site of the present South Wimbledon underground station] until the widenings were completed, the Town Council have followed the wisest and safest course. It would have been vexatious, now the work is getting so near completion, to have denied the privilege to use the Worple Road section, but the reasons against running to the Grove were sound and cogent, and the decision seems to us perfectly fair and reasonable. The Tramway Co. lost no time in availing themselves of the permission, for, on Thursday morning, the trams were running, and their comings and goings were eagerly watched at the Hill Road end of Worple Road by curious crowds, who gazed for all the world as if they had never seen a tram before. . . .'

Colonel Yorke passed the Wimbledon Hill Road–Merton–Tooting section and the Merton–Summerstown branch on June 25th and these, with the return single-track loop from Hill Road via St Georges Road and Francis Grove to avoid reversing Wimbledon-only cars, were opened on June 27th.

In April 1904 the county boundary had been altered to run alongside the northern side of the Wimbledon–Tooting railway, so that by the time the LUT arrived at Tooting it found itself owner of 246 feet of track inside the County of London, separated by a few yards from the new LCC railhead.

From the start the LUT had been anxious to erect a depot in Surrey if only to reduce the unprofitable mileage involved in running from Wimbledon and Tooting to Fulwell depot. In November 1905 it was seeking to buy a piece of land which included part of the recreation ground near the southern end of Haydons Road, on the Merton–Summerstown route.

These proposals were abortive. There have been suggestions that a small depot was in fact erected in High Street, Colliers Wood, possibly on the site just north of Walpole Road now occupied by A. Randow Limited. There does not seem to be any firm evidence to support such a theory; indeed details kindly supplied by that company dispel any such idea. In any case the area concerned is not large enough to permit stabling more than a very few cars and no building in the district bears any resemblance to the substantial type of depot which was favoured by the LUT.

In June 1907 Wimbledon Council complained that trams serving the town were carrying advertisements, contrary to agreement, though the matter was not pressed. It took strong action to enforce an onerous section of the LUT's 1902 Act by which the company was required to wood-pave the whole width of the roadway opposite to, and for 10 yards on either side of, all churches, chapels, public schools and buildings habitually used for public meetings and assemblies which fronted the tramway in the borough. Robinson told Colonel Yorke that Surrey County Council had not requested them to undertake such work.

At the directors' meeting on May 7, 1907, the solicitor reported that because of strong opposition by the Metropolitan Railway, London County Council and Hammersmith Council they had withdrawn from their Bill the powers for a proposed pedestrian subway at Hammersmith. Robinson put forward plans for a site in Plough Lane, Wimbledon, for a depot to house twenty-five cars. It was resolved to ask him to conclude the purchase within a price named. At the following meeting Robinson reported that because of the exorbitant sum being asked for the Plough Lane site he was proposing another in West Barnes Lane, Raynes Park, on land owned by the Metropolitan Water Board. Although he was authorized to acquire, nothing was done.

It was at the May 1907 meeting that Robinson had a less agreeable task to perform when he intimated that Clifton Robinson junior had tendered his resignation from the post of Traffic Superintendent. The resignation was accepted with regret. It seems clear that the young man had not, alas, maintained his early promise and had adopted an extravagant mode of life which was to lead him to the Bankruptcy Court. So far as is known, he was never again in employment, though married, but was supported by his parents. He died in his early forties. Reuben Cramp Rogers took his place as Traffic Superintendent of the LUT.

The May 4, 1907 issue of the *Richmond & Twickenham Times* featured an interesting plan by E. L. Partridge, MSA, FSI, for a new bridge and streets and tramways in Richmond. The author confessed he had been against electric trams in Richmond but had changed his mind on the completion of the Kingston lines and their attraction of trade to that town. 'Whatever may be urged against trams,' he wrote, 'they are preferable, from many points of view, to motor buses, with their nerve-racking noises, terrible vibration, breaking-up of roads and non-contribution to rates.'

Partridge thought that a Richmond–Ham tramway link might be formed by driving a new road across Petersham Meadows to avoid the S-bend through Petersham. He proposed a new street from the George Street–Hill Street junction to Richmond Road, St Margarets, which would include a widened Water Lane and a new river bridge 55 feet wide in continuation, as well as new streets and widenings in the town centre, all suitable for tramways. His plans may at least have moved Richmond Council sufficiently for them to discuss in June a new project for a municipal system.

The noise question came up once more during the year. In July

Rounding Off

Ealing Council invited other local authorities to attend a conference. In September, Hammersmith, Ealing, Hampton Wick, Hanwell, Teddington and Twickenham Councils asked the President of the Board of Trade to receive a deputation on the subject. It was said that some time before Teddington had asked the Board of Trade to inspect the track in the town under the provisions of Section 27 of the LUT's Act. The Board of Trade had not then been disposed to act unless action had first been taken against the LUT under the first part of that section.

'Lady Robinson's garden party and river fete' for the wives and children of LUT men in August 1907 seems to have excelled itself for what had come to be an eagerly awaited annual treat. About 3,000 came in thirty-four special cars from various parts of the system. One car from Hanwell carried the band attached to Hanwell depot which played on the way to Hampton. In the grounds of Garrick Villa, a huge marquee had been set up in which ninety waiters from Harrods supplied to eighty-four tables.

After lunch the company was entertained by bands from Hanwell and Fulwell and 'serenaders' until Sir Clifton and Lady Robinson arrived. After a speech, Sir Clifton, as was his custom, asked if any twins had been born since the previous year's party. There was only one set, born to Driver Mabb of Chiswick. Sir Clifton nursed the babies and gave Mabb a sovereign.

The guests roamed the grounds and went down to the riverside lawns by the tunnel under the road. On the river, some of the more adventurous LUT spirits staged a 'farcical regatta' with much sailing and punting and other water sports. Lady Robinson then presented each child with a box of Fuller's chocolates.

Every year the depots keenly competed to present the most beautiful floral device to Lady Robinson. That year Chiswick won the contest with a floral motor car.

Robinson made it a point never to be away at Bank Holiday times but remained on the spot or at least within call to encourage the staff. On Bank Holiday 1907 the LUT carried 400,000 passengers without incident and without one man having to be reprimanded. Cars earned an average of 2s a mile run.

Both Twickenham and Heston-Isleworth opposed the LUT's Bill in the 1907 session for extension of time for the Baber Bridge–Staines line. Other parts of the Bill related to: postponement of the period of purchase by Hammersmith, a matter due to arise in the year; powers for mutual LUT–LCC through running; and provisions to remove the restriction on the use of overhead trolley in Kew Road, whereupon Richmond decided to oppose. In April Heston-Isleworth Council was informed that powers for an extension of time for the Staines line would not be sought. The Act was gained on July 26, 1907.

Since March 1, 1906, Robinson had been receiving £2,000 a year from the LUT. At that time it had been decided to determine an agreement of April 12, 1902, and pay him £24,822 in discharge of all claims under that agreement.

During the year 1907 Tees-side Unionists asked him to become their next Parliamentary condidate. One can only speculate on the future of transport legislation had he decided to meet their wishes. Instead he felt that even he was due for a real holiday after fifteen years. He was given leave in November to make a world tour. He and Lady Robinson left for New York on the maiden voyage of the *Mauretania* on November 16th. Harris, his chief assistant at Chiswick, took temporary charge.

In January 1908 Chiswick Council supported Ealing's application for an injunction against the LUT for noise and agreed to contribute £60. Ealing's attitude did not prevent it from considering the formation of a committee to look into the possibilities of a municipal tramway or bus service in South Ealing. The borough surveyor and electrical engineer was asked to submit estimates for a 3-foot 6-inch gauge tramway on the Dolter stud contact system, to be operated by four cars.

At this time Hammersmith opposed the Bill of the West London, Barnes and Richmond Tramway Company, prompted on behalf of the Richmond Electric Light & Power Company. Richmond decided not to object to this company's proposal to use trolley in Barnes and Mortlake.

Powers for the Hammersmith Bridge–Richmond line had lapsed but the LUT decided to make a further attempt to cross Kew Bridge and a proposal was embodied in a Bill in the 1908 session. The Bill was examined in February 1908 and Richmond Council's consent was produced. Brentford had consented in 1906 but now decided to abstain. Under the terms of the consents by Surrey and Middlesex in 1904, the LUT would have to pay about £2,250 a year for a mere 3 furlongs! Richmond Council was negotiating with the LUT on the type of car to be run on an electrified Kew Road line.

In March, Hanwell, probably inspired by Ealing, was complaining about the state of the track, but Colonel Yorke was satisfied that the LUT was properly maintaining it. By June there seems to have been

Rounding Off

general agreement that track rather than cars caused the noise, but the Ealing Town Clerk nevertheless sent the Commissioner of Police a list showing the numbers of the cars that were considered to be noisy. At one council meeting a Mr Otway gave notice that at the next meeting he would move for the removal of the centre poles in the borough, adding, with possibly unconscious humour, that it seemed rather a tall order all the same.

The Kew Bridge Bill preamble was proved in May and the Bill passed the Lords on July 21st, receiving the Royal Assent on August 1st. Because of its large shareholding in the LUT the Imperial Tramways Company offered to find the capital for the bridge line and the Kew Road electrification. Cater Scott said that the company might have to face going ahead with both schemes, though it really required a breathing space to consolidate. The flood tide was to turn.

There was still hope that negotiations with the LCC for through running and the bridging of the short gap at Tooting would end satisfactorily.

Results for 1907 showed gross receipts of £345,570 and net revenue of £131,747. Gross receipts were £17,674 up (5 per cent) and passenger totals up by 6 per cent, from 3,370,699 to 58,725,980. Nevertheless the profit did not allow for any appropriation to renewals. After payment of interest on the debentures and a dividend on preference shares there was a balance of £10,260.

But for a poor summer the results would have been better. Cater Scott said that in the last six months of 1907 there were fourteen wet weekends and in the whole of the year there were twenty-four compared with only nine in 1906. The difference between fine and wet weather affected the takings of the Thames Valley lines by £1,242 in one week. Materials cost more. Also, there had been 340 cars in for overhaul, compared with 300 in 1906, hence an increase in the wage bill. On the other hand, a new mileage was coming into service and some mileage had not been fully developed.

By March 1908 Robinson and his wife were back from their world tour. He told a *Surrey Comet* reporter that constant application to business had begun to tell on even his sturdy, well-knit frame. They had travelled 30,000 miles in three months and he never felt better. They had been to America, Japan and China and come home through Suez.

Robinson said that nowhere were there such restrictions on electric tramways as in Britain. In the big American cities electric trams ran at up to 30 m.p.h. Tokyo had about the same mileage of tramways as the

London United Tramways

LUT. The LUT worked about 250 cars daily on average – 350 at most – at an average speed of 8 m.p.h. and with no standing passengers. Even so, they moved about 60,000,000 persons a year. Tokyo had an average of 1,000 cars a day – 1,250 at most. They averaged 20 m.p.h. and carried 300,000,000 a year. Yet Tokyo streets were worse than London's and just as congested.

A speed of 20 m.p.h. was perfectly safe on the LUT because of the cars' powerful brakes. To limit them to 6–12 m.p.h. was, Robinson thought, like harnessing an elephant to a wheelbarrow. Single-deck cars were the real answer. Passengers took too long to board double-deckers. The present loading restriction should go.

In April Robinson was host at Garrick Villa to fifty burgomasters and other German officials who were visiting England under the auspices of the British Municipal Society. The party went on to Hampton Court and returned from there to Shepherds Bush by tram.

Invigorated by his holiday Robinson presented an excellent address to the Tramways and Light Railways Association at its congress at the Franco-British Exhibition at the White City on July 9, 1908.

'I have practised my profession in the United States, the very cradle of tramways, have seen in operation every method of traction from "cows to conduits", and this intimate association has extended to four continents.'

After this personal summary Robinson gave a résumé of the early development of tramways from their beginnings in the United States 'where necessity evolved them when ordinary roads and streets were non-existent or impassable'.

He referred to his former enthusiasm for cable traction and said that the bold step of using cable as well as animal traction in Los Angeles had been justified. When he revisited that city the preceding winter he found a street railway system of 300 miles. The lines he had laid down had been the 'trend and direction'. 'My heart swelled to remember [sic] that I had planted the acorn and tended the sapling from which this noble trunk and these branching limbs had so magnificently developed.'

Robinson then spoke of the development of electric traction and of American electrical manufacturers who had 'endeavoured to push the system for all it was worth, and for a few years for a good deal more'.

In this country the Board of Trade's electric traction regulations of 1894 had increased constructional costs but also encouraged a type of

Rounding Off

installation and equipment unsurpassed anywhere else in the world.

The decision by the House of Lords on the price of purchase by a local authority, under Section 43 of the Tramways Act, made such bodies rush to become tramway owners. That price was the cost of the original construction less an allowance for depreciation dependent on the condition of the system on acquisition. The ruling meant that the hope of raising capital by private enterprise had almost to be abandoned. A few stronger companies came to terms with authorities.

Robinson said he was the first to use high-tension a.c. transmission in Britain for substations in order to conform on an extended line with the Board of Trade 'seven-volt drop' restriction – the reference was to the Dublin Southern.

Why did the struggle to electrify the LUT go on for seven years? One reason was the attitude of the LCC. When, before the LCC, he advocated using trolley at Shepherds Bush he was threatened, in effect, that people would tear up the posts. He had perforce to submit to conduit in Hammersmith but when the local council realized its streets were to be used for experiment, resolutions were passed and the conduit clause was rescinded in the next session. The LCC was so long anti-trolley, alleged Robinson, that many borough councils followed suit and logically vetoed overhead.

Even though the LUT had had to negotiate with thirty public authorities, whose price of consent for road widenings, capitalized wayleaves and so on aggregated £1,000,000, it had earned on an average 5 per cent in its six years of electric working and carried 300,000,000 people.

The Light Railway Act made for a quicker and rather cheaper procedure than by a Bill. There was no veto by authorities to contend with, but on the other hand a railway might veto on the grounds of competition. Also, a light railway Order was not granted for a line wholly within the area of a single authority.

A step forward to mitigate the effects of veto had been taken when both Houses disallowed the vetoes of the Middlesex and Surrey county councils and allowed the LUT's Kew Bridge link to go forward.

Robinson reiterated his belief that double-deck cars generally caused delay in loading but the no-standing rule made them necessary.

The LUT ran some sixty workmen's cars daily at a fare of less than $\frac{1}{4}$d a mile. The average fare over all the system for ordinary passengers was less than $\frac{1}{2}$d a mile.

In September 1908 the LUT asked the Board of Trade to sanction a speed increase from 8 to 12 m.p.h. in Kingston Road, Teddington, a reasonable request as the road was quite wide and straight.

About this time Twickenham was being difficult. In September the council complained of noise to the Commissioner of Police, who replied rather ambiguously that if the track was not corrugated there was no noise that could not be prevented. Back came the council in November with complaints about bad track in places, overcrowding at times and unpunctuality.

Robinson retorted sharply. 'Many of the criticisms were the outcome of complaints based upon a blend of insufficient knowledge and prejudice, stimulated occasionally by the assumption that the company are not cognisant with the elementary principles of their own business.'

Twickenham would not be put off and agreed with Ealing to request the Board of Trade to abate the noise of cars and occasionally inspect the state of cars and track. The LUT had already asked the Board of Trade to name an arbiter in its differences with Hammersmith over the state of track there.

In June 1909 Wimbledon decided to oppose the LUT's application to the Board of Trade to raise the maximum speed allowed over part of the system from 14 to 16 m.p.h. Acton, which enjoined a minimum of 8 m.p.h. in some places, opposed an increase between Birkbeck Road and Gunnersbury Lane.

In June Robinson told Twickenham that it was the only council to complain of noisy cars! The Council then canvassed other authorities. Hampton replied that it had often complained, though not since August 1908 as the company treated complaints cavalierly – the council now wrote only to the Commissioner of Police. Teddington replied that it had asked the police several times not to license noisy cars. At this point, Merton & Morden stepped in and decided to complain of noise to the Board of Trade.

Not all the complaints can have been ill-founded. In its issue of September 22, 1910, the *Electrical Times*, certainly no enemy to the LUT, wrote: 'The greatest offender known to us is the London United, whose running in certain parts of Surrey and Middlesex has engendered an offensive habit of yelling all along the route. In the absence of a tramcar it is most embarrassing to ask a Twickenham or a Kingston chemist for drugs of a personal nature unless the street is cleared of people.'

Chapter 11

Chill Winds – And New Brooms

The report and accounts for 1908 made unhappy reading. The traction expenses were up, because of the increased price of coal and stores, but there was a small saving on traffic costs. The cost of maintenance and repairs had risen from £48,676 to £51,492 but Cater Scott said that showed they were doing their best to keep the track in order even in bad times. He blamed the want of traffic development on bad weather, the unexpectedly poor results of travel to and from the Franco-British Exhibition, and bus competition.

There were 129 motor buses on routes converging on Shepherd's Bush, said Cater Scott. They did not carry enough passengers to pay but they diverted traffic from the trams. Buses had no compulsory stops. They nursed the trams and took away their traffic. True, bus companies paid rates on their garages, but tramways paid on their total income and did not receive the three-quarter reduction granted to railways.

Such burdens were accepted, went on Cater Scott, when the company first went to Parliament. With the advent of motor buses the situation was quite altered. Any man who attempted to build a tramway today under the conditions imposed by the Acts would be insane. Under such conditions, tramway building, save by municipalities, was finished. The remedy in London was a Traffic Board to control all traffic.

Twenty-five per cent of the company's capital had gone on street widenings. Cater Scott allowed that they enjoyed a rich summer traffic to the Thames Valley resorts but the winter traffic was hardly enough for economical running. Until the company could put by sufficient year by year for renewals, the preference shareholders would have to be content with 2½ per cent. It was difficult to forecast when they could return to a full preference dividend.

A Mr Fitch remarked at the meeting that the Ealing section was unsatisfactory and ill-managed. The conductors were insulting. As for the Uxbridge extension, it was money thrown away. But motorbuses were at their last gasp and would sooner or later close up. [The day of the General's reliable B-type bus was not far distant!]

Cater Scott replied that they hoped the Uxbridge line would develop. [Even here the LUT was unlucky, for a building recession had made itself felt after the first rash of building in Southall and elsewhere.] He went on to say, significantly, that they would have hesitated to build some lines if they had foreseen bus competition. He did not understand why the Tooting gap had not been closed. If the South London trams had been in the hands of a company, the LCC would have been the first to demand such a connection. Powers for through running were included in the company's present Bill. In any case the LCC was to fit trolley arms to some cars. The LUT would fit plough carriers to some cars to enable them to use the LCC conduit. If the traffic was unbalanced, the matter could be arbitrated. (This part of the Bill sought reciprocal running powers, the LUT to be allowed to run over the LCC from Summerstown to Victoria and to the Embankment via Wandsworth, and from Tooting to the Embankment.)

A. L. Coventry Fell, the LCCT manager – and a notable one even if rash enough to predict the extinction of the motorbus – considered that the extra trouble of running LCC cars beyond the boundary would be inadequately rewarded. LCC trams averaged £187 a mile annually, the LUT a mere £55. When the journey was more than one hour the train successfully competed. [The LBSCR began its 'Elevated Electric' service on its South London line during the year and soon made deep inroads into LCC traffic in that area.]

The LUT argued that the purchase of its Hammersmith lines by the LCC would dislocate its traffic and end through bookings with the Underground. The Bill, therefore, also asked for a postponement until 1919 for the 'Young's Corner' lines and until 1924 for the Askew Road line, to conform to the dates of the Chiswick and Acton purchase respectively. It was rejected.

Robinson's hopes of building a depot in or near Wimbledon were again dashed in October 1909. He had found a site for a depot to hold thirty cars, at an estimated cost of £8,500 for ground and buildings. Since the Surrey lines had opened it had cost the company £2,061 in dead car mileage to and from Fulwell. The expenses in connection with a new depot would be £1,252. It is a pointer to the state of the undertaking at this time that Robinson was told that, although such a depot would be most desirable, they could not afford it. This would seem to have been

Chill Winds – And New Brooms

Robinson's last major effort for the LUT. One can imagine how the change from a bullish to a bearish policy must have galled his ebullient nature.

In November Heston-Isleworth, Southall-Norwood and Twickenham councils decided to send a deputation to Middlesex Council to press for a tramway or light railway between Southall and Twickenham via Norwood Green, Hounslow and Whitton, a route over much of which both the LUT and Middlesex had projected lines a few years earlier. Twickenham, though, seems to have been a rather reluctant signatory, observing churlishly that 'tramways had not, up to the present, resulted in any good to Twickenham'.

On October 8, 1909, the MET at last opened the Harlesden–Acton route. There was no physical connection at Acton but the LUT gained a new and useful interchange which placed it in communication with the whole of north-west and north London, in which the MET had all but completed its far-flung system, mainly as a lessee of Middlesex County Council.

On January 14, 1910, Robinson wrote to Cater Scott: 'Referring to my interview with Sir Edgar Speyer and yourself, I now write to formally resign my appointment as Managing Director and Engineer, and my seat on the Board of the London United Tramways Company. I shall be pleased to consult the convenience of the Board as to the date of carrying out the contemplated change, and to afford to Mr Stanley every assistance on his taking over the management.'

'Mr Stanley' was Albert Henry Stanley, born at Derby on November 8, 1874. At the age of eleven he had gone with his parents to the United States, where his father took up an appointment in the Pullman works at Detroit. Young Stanley determined to enter the service of the Detroit City Street Railways, which at that time operated 43 miles of horse-car lines. He began as a messenger, then became a superintendent and by the time he was twenty he controlled the whole system.

In 1898 Stanley served as an ordinary seaman in the United States Navy in the Spanish-American War. When he relinquished the post of General Superintendent at Detroit in 1903 the system had grown to one of 550 miles under electric traction.

Stanley had continued his education by taking courses in transport and engineering at Detroit technical institutes. In October 1903 he was appointed Assistant General Manager of the Street Railway Department of the Public Service Corporation of New Jersey. In January 1907 he became General Manager of the corporation, but in April returned to

England to become General Manager of the Metropolitan District Railway and the London Electric Railway (Bakerloo, Hampstead and Piccadilly tubes). As the youngest General Manager of a British railway, Stanley soon made his mark on London transport, with the aim of bringing all services, both rail and road, under unified management.

Accepting Robinson's resignation with much regret, the board agreed to pay him his salary for one year from March 1st and desired to 'express its recognition of the unsparing efforts of Sir Clifton Robinson to secure the best interests of the London United Tramways, the great ability he had always shewn during the many years devoted to their promotion, construction and operation, and the courtesy and kindness which all the Board have always experienced at his hands, and also to assure Sir Clifton Robinson that in his retirement he will carry with him the sincere wishes of his colleagues on the Board for his future happiness and welfare'.

It was agreed to release Robinson before March 1st as he wished to go abroad as soon as possible and Stanley was ready to take up his post at once. Robinson had accepted a commission from Speyer Brothers to look into traction possibilities in the Philippines. He told an agency representative that police regulations made in coaching days still hampered British tramways. Unlike railways, they could not carry more passengers than they could seat. They had to buy the assent of local bodies and submit to blackmail. The formation of the Traffic Department of the Board of Trade was a step towards dealing with the problem of London traffic. But tramways and railways were limited in Britain and private enterprise was being snuffed out. There was no more utility for more electric undertakings. But abroad there was scope for British capital.

Robinson left London on February 8th with Lady Robinson and their son to join a Norddeutscher Lloyd ship at Genoa and travel via Hong Kong. He told the *Middlesex County Times* that 'The result may be the opening up of fresh enterprises for British capital in the Far East. I shall be away several months as my observations of the different systems and possibilities in Oriental countries will have to be made in thorough manner. . . . Mr Stanley will find the United Tramways in an excellent condition from every point of view. The tramways will really form an integral part of the underground railways and will be managed from the offices at St James's Park Station, where Mr Stanley is located.' Robinson added that he would remain on the boards of the UER and MDR.

Just before he left England, Robinson gave a dinner, on February 2nd,

Left: The Mayor of Kingston prepares to take type T No. 320 across Kingston Bridge in March 1906 as the first electric tramcar to cross a Thames bridge. Robinson stands next to car on right-hand side. Clifton Robinson junior with one foot on the footboard. (*Courtesy Kingston Library*)

Centre: Bedecked cars cross Kingston Bridge to inaugurate the opening of the Surrey lines in 1906.

Bottom: The Mayor of Kingston drives No. 320 on the inaugural tour of Kingston lines in March 1906. Robinson stands on the footboard. The military gentleman behind the destination indicator is probably Lt.-Colonel Yorke, the Board of Trade inspector. (*Courtesy Kingston Library*)

Top: Bruce, a coloured driver, at the controls of type T No. 321 in Claremont Road, Surbiton, during the Board of Trade inspection in March 1906. (*Courtesy B. Woodriff.*) *Centre:* Type T No. 330, heading for Kingston Hill in 1906, passes the site of the present-day C & A store in Kingston. *Bottom:* Type T No. 309 on Dittons–Richmond Bridge service at Hampton Wick in 1906. (*Courtesy B. Woodriff*)

'Tobruk Taken' reads one of the placards, but the year is 1911, as No. 168 of type W leaves Surbiton for Kingston Hill. (*Courtesy B. Woodriff*)

In Victoria Road, Surbiton.

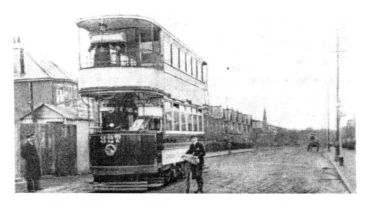

Type T No. 327 at Tolworth terminus soon after the opening of the line in 1906.

Left: Car on King's Road–Ham Boundary shuttle service. *Centre:* Type W No. 162 about to turn from Portsmouth Road into Brighton Road, Surbiton. *Bottom:* Dittons terminus, with type T No. 312 in foreground.

Chill Winds – And New Brooms

at the Trocadero to the Metropolitan Association of Electric Tramway Managers, of which he was chairman. In announcing his retirement from the LUT he graciously referred to Stanley as one who had won his spurs in tramway work in the United States. Mr H. E. Blain, General Manager of West Ham Corporation Tramways, responding to Robinson's toast of the Association, said that Robinson's limit of activity had hardly been begun. All tramway managers in London owed him gratitude for his fight in the Commons for through running. They were indebted to him that evening for the presence of Sir Herbert Jekyll, for most of them were looking forward to the work to which he believed Sir Herbert had set his hand, the establishment of a traffic board for London, which would unite or co-ordinate the tramways.

As chairman of the UER, Sir Edgar Speyer presented Robinson with a pearl scarf pin as a souvenir of the work he had done as a pioneer of electric traction in London. In his reply Robinson recalled his early days with the West Metropolitan and then said that since the Royal Commission nothing had been done for London traffic save what Sir Herbert Jekyll had done to keep alive the light of the Commission.

Albert Stanley told the audience that Robinson had declared to him that, as far as he knew, the LUT staff was practically unequalled anywhere.

In this euphoric atmosphere, Robinson sailed away to the East, leaving the new brooms of Electric Railway House ready to sweep. Changes were not long in coming. Reuben Rogers was one who then left the LUT service, exchanging electric traction for the motorcar business.

At the directors' meeting in April Stanley reported the appointment of Zac Ellis Knapp to be his assistant on the LUT at Chiswick, at a salary of £1,000 a year, of which the LUT would pay £750. Knapp, a North Carolina man, had been for some time assistant to the chief engineer of the Underground. He had made his name in building power houses in Chicago. As Assistant Manager of the LUT, Knapp would accomplish much in three difficult years. He would tighten up organization, introduce new methods of staff discipline and grant annual leave to tram crews and maintenance staff. (He later became chief engineer of the Underground and died in 1926.)

At the annual general meeting on April 22nd Cater Scott spoke of the serious drop in receipts since 1906. In 1909 both the bad weather and bus competition had hit their pleasure traffic hard. On Sundays the buses varied their routes to invade country roads and offer still more competition. The electrified Metropolitan and District railways were efficient competitors.

For some years they would have to face large outlays for depreciation and use much surplus revenue for maintenance. They also had to meet heavy claims for compensation and had had to seek a large bank loan.

Receipts for 1909 were £318,226 gross and expenses were £226,566. After payment of interest and provision for tax there was a balance of £21,950 and £21,000 would be placed to the renewals fund. Gross receipts were £30,163 down and working expenses were £6,334 up. The increased costs arose largely from extra repair work on cars and track, but there was a saving in current because of the reduced price of coal. It was advisable to carry out certain repairs to cars and track at once.

All this does not quite seem to add up to the clean bill of health that Robinson said Stanley would find!

Cater Scott told the meeting that they had sought a closer contact with the Underground. Robinson could have put obstacles in their way, but he had helped them and his retirement was a friendly act.

At the Trocadero dinner Balfour Browne had accused local authorities of blackmail. His allegations were followed up by a *Surrey Comet* reporter who interviewed members of councils in the LUT area.

The Mayor of Kingston said that it had been a matter of business on both sides. But had it not been for the LUT Kingston would never have had such a thoroughfare as Clarence Street now was. He thought the LUT had suffered more from the extortions of private persons. As districts developed, tramways would become paying concerns. If the LUT could not pay its way, how could a Kingston local tramway have paid? He was glad that he had opposed such a project.

The chairman of Surbiton council said that the Kingston–Surbiton horse bus had charged a 2d fare, so no wonder there had been a plebiscite in favour of the LUT. He did not consider a wayleave was blackmail. What about reduced property values and the increase in empty houses, as in Claremont Road?

Malden council's chairman was hostile to Robinson. He did not regard the LUT as British in either method or manner and only regretted that local authorities could not prevent promoters from getting as much as they did. 'The distinguished champion of private enterprise (and, I presume, profit) who has left us, always seemed to me to be unable to realize that the assets of the street franchise merchant on the other side of the Atlantic are not suitable to British conditions.'

Alderman Hamshaw of Wimbledon said that his council had had the experience of neighbouring authorities to guide them in tramway matters. Legitimate protection was a '*quid pro quo*', 'to use one of Sir Clifton's happy phrases'.

Little Hampton Wick was generous. The vice-chairman of the council said that the three widenings in Hampton Wick had been a spontaneous act by the LUT. There was never friction with Robinson, who had treated them handsomely in all transactions.

Mr W. G. Collier, who had been chairman of the Tramway Committee of Teddington Council when the LUT was laid, thought Robinson had had the best of the bargain, for there was still an important widening between Udney Park Road and Field Lane unfinished. Had his council pressed, the LUT would have had to buy up seven or eight shops. The new railway bridge went far beyond the council's expectations but a new bridge had been needed in any case. [Collier did not say when it might otherwise have materialized!] Tramways were a mixed blessing to the town. The service was irregular and unsatisfactory and property values had dropped 25 per cent along the route. The council had not insisted on double track throughout, with a 9-foot 6-inch margin.

Mr William Poupart, for some years Chairman of the Highways Committee of Twickenham council, told the *Surrey Comet* that every widening in the town had been done after amicable conferences with Robinson. Two widenings remained to be carried out: Cole's Bridge and part of Heath Road, where the LUT was committed to half the cost if the work was ever done.

In July 1910 Ealing Council's surveyor reported that the welding of tramway rail joints was proceeding satisfactorily. He believed it would tend to lessen noise. Mr Stanley had promised that the overhaul and repair of cars would be done as rapidly as possible.

In November the LUT told Heston-Isleworth that it was welding joints on the Uxbridge route at a cost of £2 per joint. The Hounslow route was next to be tackled. The company added the information that 140 – an exaggeration – of the open-top cars were to be covered, at a rate of five a week.

Chapter 12

A 'Tramway King' Passes

Not long after the Robinsons' return from the East – Lady Robinson later described Manila as the most delightful place she had visited in her foreign travels – Sir Clifton received a commission from a financial house to investigate and report on the possibilities of commercial development in Newfoundland.

Before he and Lady Robinson left England again, he once more sat down to dine, as their president, with his fellow members of the Metropolitan Association of Electric Tramway Managers. On this occasion he was being entertained by them, in commemoration of the jubilee of tramways in Britain.

Sir Clifton, declared Blair of West Ham, had said that he had handled every species of transport from cow to conduit. Blair gave the route mileage of British Tramways as 2,526, laid at a capital cost of over £71,000,000.

Robinson was in great form in recalling Train. He had a photograph of the first tramcar in Birkenhead, which was crowded with celebrities. Train had made one of his 'perfervid orations of the eloquence and magnetism of which this generation can have little conception'. Robinson did not believe that he had ever heard a man who could use language with such strength, such terseness, such force as Train. He remembered the first day he ever saw a tramcar and he also remembered his first interview with Train. Train was one of the few Americans he had met who never drank strong liquor, who never used strong language – he was sorry he had not taken after him – who never used tobacco, who seldom did anything which any average man would do.

Train, continued Robinson, was an extraordinarily handsome man who had married a beautiful lady and taken her to Australia. So that his son could be born under the Stars and Stripes he fitted out a special clipper ship. When a girl was born 'Train was near breaking forth'.

After a reference to the opening of Train's Marble Arch–Porchester

A 'Tramway King' Passes

Terrace, Victoria Street and Westminster Bridge–Kennington lines and to the American-built Aldgate–Bow and Moorgate–Angel lines of 1871, Robinson looked ahead. It was difficult to say what would take place in the next fifty years. He did not know whether aerial navigation would play a part in rapid transit but cheap fares and rapid transit were essential to social life.

While in Newfoundland Robinson was asked by Bull (now Sir William), acting for the depositors of the Charing Cross Bank, to report on the Atlantic, Quebec and Western Railway, of Canada, in which the bank had a large holding.

From his Canadian assignment, Sir Clifton, with Lady Robinson, went south to New York, intending to sail for home on November 16th.

On November 6th, the Robinsons attended a dinner party in New York given by Lady Robinson's brother Shaw Martin. While they were in a tramcar on their way back to their hotel, Robinson complained of illness and collapsed. He was taken to a drug store near, but died in a few minutes.

He might well have chosen his end, in a tram in the greatest city of a land that was almost a second home to him and almost certainly was his first spiritual home, a city which had seen the first of all street railways and had taught him the basis of his craft.

The Times obituary notice recalled that Robinson had failed in his attempt to gain running powers over the LCC system. He had urged that tramcars should be allowed to travel faster and declared that the need for double-deck vehicles arose because of limitations on the use of trailers and on the total number of passengers permitted to be carried, but they caused delay in boarding and alighting.

Robinson had been optimistic, continued *The Times*, about British tramway enterprise, as witness his paper 'The Tramways of the World' to the Tramway & Light Railway Association, of which he was a Vice-President – 'I see no reason why they should not enjoy an indefinite period of financial prosperity if skilfully managed and conservatively financed. It is true that many of us groan under imposts of local authorities, not only in huge exactions, but in other vexatious directions.' Robinson hoped that the public, 'having tasted the sweets of rapid transit in urban streets, would give Parliament "moving orders" to substitute something for Standing Orders that throttle business, indulging class or special legislation and conferring upon certain vested interests a monopoly of roads which should be equally at the service of all'.

Most of the other obituary notices in the general press stressed Robinson's romantic career, which had seemed the epitome of Victorian self-help.

The *Evening News* of November 7th said: 'A former colleague of Sir Clifton in the control of the London United Tramways described him as a "man of no hobbies, except traction". He watched the progress of aviation with keen zest, and travelled considerable distances in order to attend flying meetings. In the aeroplane he recognized a future means of transit for the people, and possibly something of a rival to tramways'.

The *Railway Gazette* believed that 'Sir Clifton may, without undue exaggeration, be termed the "Tramway Stephenson", for, like George Stephenson, he played a leading part in helping to place on a commercial basis a new means of transportation that other men had originally devised but lacked the necessary organizing powers and pertinacity to establish on a sound footing. . . . He will rank as an exponent of the ideal type of transportation man, for singleness of purpose and of interests were exemplified in his career to a remarkable extent. He talked, thought and dreamt tramways'.

The *Light Railway & Tramway Journal*, in its November 11th issue, said: 'When Sir Clifton Robinson, on February 2nd last, in that pathetic speech which he described as his "swan song" announced his retirement from the managing directorship of the London United Tramways, those present on that occasion little thought that before the year was out the tramcars on that system, of which, amongst others, he was the founder, would by flying Union Jacks half mast high on the trolleypoles. . . .'

In its December 2nd issue the same journal wrote: 'Robinson's career showed how it was possible to mount from the very bottom to the very top. Robinson was not without some faults of temperament and was sometimes brusque, but he had a kind and generous heart and never knew when he was beaten. His career exemplified pluck and perseverance.'

A businesslike note, which Robinson might have approved, was struck by the *Electrical Review* which said: 'His regretted death removed a familiar figure from the giants of the tramway world, and had it occurred some years ago, the effect on some of the Stock Exchange prices would have been considerable. But latterly he had not been much concerned with the practical management of the companies with which his name was most closely identified. As it was, the LUT's preference shares fell $\frac{1}{4}$d on the news of his death.

A 'Tramway King' Passes

'He recognized the power of the lay press', continued the same journal, 'and the lay press was ever ready to make use of him as a subject upon whom, and whose views, the popular reader loved to dwell. . . . He could be genial or imperious by turns. . . . His life is one more illustration of the truth that men can "get on" if they possess grit, will, energy and persistency, provided the opportunity turns up – or they turn it up – and they grasp it.'

Nor were the journals in provincial towns where Robinson's tramways ran less forthcoming with copy. The *North-Eastern Daily Gazette* wrote: 'A smart, far-seeing, shrewd man of business, Sir Clifton was always eager to entertain a business "proposition", but he found the Middlesbrough Town Council an exceedingly difficult nut to crack. There has been many a battle royal between Sir Clifton and the Middlesbrough Corporation, in which victory has by no means always rested with the civic authority. Indeed, one councillor, half humourously, half despairingly remarked "Clifton Robinson is clever enough for anything". But in spite of all the controversies between himself and the Corporation, Sir Clifton's personal relations with all whom he came in contact with on Tees-side, were of the kindliest and best.'

Robinson's body reached England on November 17th. At a requiem Mass held next day at the Church of the Immaculate Conception, Farm Street, Father Bernard Vaughan spoke of Robinson's scrupulous integrity and high sense of honour. Crowds lined the route to St Mary's Cemetery, Kensal Green. Inside the cemetery 400 LUT officers and employees lined the drive. At noon the whole LUT system stopped for one minute.

In addition to members of the family, the mourners included members of the aristocracy and transport personalities. The Japanese Ambassador was present. Others included Lady Malcolm of Poltalloch; Lord George Hamilton; Catherine Lady Decies; Lady Lawrence; Sir John and Lady Roper Parkington; Sir Albert and Lady de Rutzen; Sir Thomas Pile; Sir T. Brooke-Hitchings; Sir James Szlumper; Professor Silvanus Thompson; Sir George White; Sir Edgar Speyer; Albert Stanley; W. E. Hammond, Traffic Manager of the MET; J. K. Bruce, Traffic Manager of the LCC tramways; Z. E. Knapp; W. J. Verdon Smith; A. Mason, SMET Manager; R. Graham, Secretary of the Central London Railway; Mr Sire, representing the Northern Railway of France; and A. de Twickham, Secretary of the Tramways & Light Railways Association.

The presence of Henry Wood testified to Robinson's love of music and that of A. Coke of Our Dumb Friends League to his love of animals, shared by Lady Robinson.

Robinson lies in the northern part of St Mary's Cemetery, close to the wall which separates it from the London Midland Region main-line railway to the north-west. Even though all tramways have gone from London, trains of the London Midland Region of British Railways, operated on an electrical system of which Robinson would certainly have approved, thunder past. An impressive carved Celtic cross bears the inscription: 'In memory of Sir James Clifton Robinson, JP, Civil Engineer, 1 January 1848 – 6 November 1910.'

At the meeting of LUT directors held on December 7th, with Cater Scott in the chair, it was resolved 'That the Board desire to place on record their deep regret at the sudden death on the 6th November of their late colleague and former Managing Director, Sir Clifton Robinson, and to express their high appreciation of the valuable services rendered by him. The Board further desire to express their sympathy with the widow and son of the late Sir Clifton Robinson'.

Probate of Robinson's Will was granted to Sir George and Samuel White, as trustees and executors. The Will confirmed three settlements made in his lifetime to Lady Robinson and their son. The residue of the estate was £13,641 6s 9d gross, left on similar trusts to those of his principal estates included in settlements. Personal estate was £2,516. Household and personal effects, horses, carriages, motor cars and consumable stores went to Lady Robinson and the residue on trust to Clifton Robinson and his wife and issue, and, failing issue, to Clifton Robinson, with the remainder to Mrs Clifton Robinson and the ultimate remainder to Lady Robinson.

Chapter 13

LUT, MET and BET

There was soon more trouble with local authorities over track. At a meeting of Ealing Council in February 1911 a letter from Middlesex County Council was read. It said that unless track was put into proper repair the county surveyor and engineer could not certify, when the accounts were rendered, that the track had been properly maintained and repaired. Middlesex recommended that the LUT be notified that if the defects were not remedied it would have to consider taking action under Section 28 of the Tramways Act.

The LUT had already told Ealing that it was trying to prevent grinding when cars turned corners, and that cars were being overhauled at both LUT and Underground shops to speed the work.

By April Hammersmith had asked the LCC to help to secure the reconstruction of track in Uxbridge Road, but the LCC could not comply as it had not yet come into possession of the Hammersmith lines. For three years Hammersmith Council had been serving notices on the LUT about the state of track and had at last decided to apply to the Board of Trade to name a referee. The Council would not oppose a junction between the LUT and LCCT once the LCC had taken possession.

Agreement was reached in May on interchangeable tickets between the LUT, Underground and LGOC and a joint committee was formed to settle details.

Whitsun traffic was heavy. Takings on Whit Monday were almost £3,600 and about 440,000 passengers were carried. The total was more than in 1910 but not quite that of Whit Monday 1908.

The year 1911 was remarkable for trolleybus schemes, following the success of the trolleybus inaugurations at Leeds and Bradford on June 20th.

As early as January 10th the Chairman of Chiswick UDC reported that Christopher John Spencer, General Manager of Bradford Corporation Tramways, had been asked to advise on the suitability of a proposed trolleybus scheme for Chiswick. Spencer reported favourably, advising that it would be preferable to either a tram or bus route.

The route, 2 miles 31 chains in length, was similar to that for which the LUT, Chiswick Council and Middlesex Council had all tried to promote a tramway some years before. It ran from the High Road along Chiswick Lane, Mawson Lane, the south end of Devonshire Road, Burlington Lane, the south end of Sutton Court Road, Fauconberg Road and Sutton Lane back to the High Road. Vehicles were estimated to cost £5,000 and electrical equipment £6,000.

Chiswick Council published the Bill on November 17th, 1910. When the scheme received the Royal Assent on August 18, 1911, Chiswick became the fourth British authority to gain trolleybus powers. The Commissioner of Police doubted whether trolleybuses were suitable for a London suburb.

Powers were to become effective, subject to BOT consent, after January 1, 1913. They included authority to lease to the LUT, a provision both parties agreed on in 1913. They were never exercised. London would have to wait another twenty-one years – Chiswick in particular another twenty-five – for trolleybuses in public service. Before that Spencer would be intimately concerned with Chiswick and its transport.

Heston-Isleworth Council promoted the Southall, Hounslow and Twickenham Railless Traction Company as a municipal venture. The Bill, published on December 18, 1911, shows that the route was from Southall Town Hall via Norwood Green Road, Lampton Road, Whitton Road and Kneller Road to Twickenham Town Hall. A branch had been considered from Lampton to Richmond Bridge via Spring Grove Road, St John's Road and St Margaret's Road.

Both the LUT and District Railway petitioned against the Bill, which failed on Standing Orders.

In September 1911 Ealing Chamber of Commerce sent a deputation to study trolleybus operation in Leeds and Bradford. A route was proposed from Ealing Broadway via South Ealing Road, Little Ealing Lane, Northfields Avenue, West Ealing, Argyle Road, Pitshanger Lane and Eaton Rise. Ealing Council's surveyor and electrical engineer reported on the scheme, which was rejected in March 1912.

LUT revenue in 1911 was £125,768. The balance was up by £25,618 to £56,432 and the number of passengers carried by 2,414,677 to 62,547,128.

At the annual general meeting Cater Scott said that although results

LUT, MET and BET

were better than in 1910 they could still not declare a preference dividend. They were working up from the low water of 1909. Receipts, at £343,987, were £10,328 up on 1910 and £25,761 up on 1909. Fine weather – it had been a glorious summer – had helped them but it had also helped the buses, whose competition was keener than ever. Working expenses were up by £4,888, of which wage rises accounted for £2,414. The staff now enjoyed seven days' annual leave with full pay and at least part of every Sunday free.

The track, reported the chairman, was in better state than for some years. They had repaired some 11,250 joints by cast welding and smoother running had resulted.

In November 1911 Colonel Yorke, normally sympathetic to the LUT, had told Brentford Council that, while he could not declare the track there a danger to traffic, he had advised the LUT to carry out desirable repairs at once. He added that where remedies had been tried in some places they proved as bad as the defects.

A total of 102 open-top cars had been top covered in 1910–11. Unfortunately, it cannot be said that the results were either elegant or comfortable. As one writer has said, it was very much of a make-do-and-mend job. The ungainly conversions, carried out at a time of financial stringency, would later do much to damn the LUT in the districts they traversed.

For a start, the upright for the trolley boom was left in position. At first, merely ends and a roof were fitted and the wrought iron sides were left. Later, new top deck sides were fitted but the window openings were not glazed. Instead there were spring blinds to pull down when it rained. Later still some of the openings were glazed. The end doors were set diagonally. All in all, a draughty and cheerless interior was produced.

After gaining two extensions of time for the Kew Bridge link (August 3, 1909 and June 2, 1911) the company at last decided to wash its hands of Richmond. At a meeting of the first mortgage debenture holders in April 1912 it was resolved to approve a proposal to abandon both the Kew Road horse line and the plans to electrify it. Cater Scott said that the company would ask for an extension of time but they still hoped to abandon, to end a loss and rid themselves of the liability of spending money on an electrification which he doubted would succeed. Richmond's insistence on conduit made financial success impossible.

By agreement with Richmond Council, the LUT kept up a twenty-minute token service from 8 a.m. to 11 p.m. using only car No. 22. For the time being two drivers, two conductors and four pairs of horses were maintained. Other staff were absorbed.

The end came soon. The *Richmond and Twickenham Times* of April 27, 1912, carried a piece entitled 'A Fond Farewell – How They Brought the Last Car from Kew to Richmond'. 'Unwept, unhonoured and unsung, the last of the trams on the Kew–Richmond line made its journey on Saturday last. No casual passengers of the common breed were allowed to snatch a last fearful joy. Nor were the winged Pegasi urged forward by a professional charioteer. Over the well-worn – so very well-worn – track the course was steered by Mr Wright, the Assistant Secretary to the LUET Company – all except the last stage, which was taken over by Mrs Wright.

'. . . whole rivers of ink have been used up, reams of paper spoiled and innumerable columns of the Press occupied with the statements and arguments of the contending forces, to say nothing of the interminable speeches that have been delivered. The practical result of it all is – a few old tramcars that the company will be willing to sell to junior sporting clubs who wish for cheap dressing rooms, and a mile and a half of useless track which is shortly to be pulled up to make way for the wood paving, thus enabling the motor buses to run with greater speed and comfort into Richmond from the very quarters of the Metropolis it was desired to guard against.'

The last public tram route ran on April 20th. It left Richmond depot at 10.40 p.m. and got back at 11.20. It was driven by W. Banfield, who had driven the first car in 1883. J. Short, foreman at the depot, had been with the company for seventeen years. He had begun under Robinson as a trace boy at Bristol in 1882, when four-horse cars were being used. He moved with Robinson to the Highgate cable line, later becoming a driver and timekeeper at Chiswick and transferring to Richmond in 1902.

The offices and stables at 125 Kew Road underwent various changes of ownership before becoming a fire station in 1932.

LGOC bus route 27 had been extended from Turnham Green to Richmond in December 1911. With the suspension of the horse car service, it was further extended to Twickenham. The horse car line was officially abandoned on May 31, 1912.

The LUT was finally rid of a profitless anachronism which had cost it £1,250 a year in rates alone. For the track and setts it received £200 from Richmond Council.

LUT, MET and BET

A far more important development took place on November 20, 1912, when the London & Suburban Traction Company was incorporated to consolidate the interests of the shareholders of the LUT, MET and the Tramways (MET) Omnibus Company, a new company by which the MET proposed to fight bus competition with buses.

The scheme had been promoted by the BET, which controlled the MET. Negotiations resulted in the MET buses being taken over by the LGOC, a member of the Underground group since January 1912.

From January 1, 1913, the LUT came under BET control. In March 1913 it became a member of the British Electrical Federation and its offices were transferred to 88 Kingsway. The Underground Electric Railways had a substantial but not a controlling interest in the London & Suburban. The South Metropolitan Electric Tramways & Lighting Company came under London & Suburban control in June 1913.

LUT net revenue for 1912 was £109,793. A sum of £41,228 (a drop of £15,203) was carried forward and £15,000 was placed to the general reserve and £25,000 to renewals and contingencies. The number of passengers carried was down by 1,407,843 to 61,139,285.

Track in Uxbridge Road had been relaid and sections in King Street and Goldhawk Road would soon be treated. The LCC was to contribute to the cost of trackwork in Hammersmith and was empowered in 1913 to work over the Askew Road line.

When the LUT, MET and Tramways (MET) Omnibus came under common management, James Devonshire (later Sir James Devonshire, KBE) became Managing Director. Albert Stanley remained on the LUT Board. A. H. Pott became Engineer and General Manager, and A. L. Barber replaced W. G. Verdon Smith as Secretary. Both were MET men. Cater Scott resigned from the LUT Board and his place as Chairman was taken by W. M. Acworth, a director of the London Electric Railways, noted railway economist and statistician and author of the classic books *The Railways of England* and *The Railways of Scotland*.

Devonshire, a director of the MET and henceforward of the Underground Group, London & Suburban, LUT and SMET, had begun his career in 1888 when he entered the service of Laing, Wharton & Down Construction Syndicate, predecessor of the British Thomson-Houston Co. Ltd, of which he became Manager and Secretary in 1891. His company's contracts included the Dublin Southern Tramways, Bristol Tramways and LUT electrifications. He later joined British Electric Traction, which provided the capital of the MET, whose managing director he became in 1902. As a director of the North Metropolitan Tramways in 1903 he negotiated the sale to the LCC of that company's lines leased by the Council in the County of London.

In 1913 Acworth reported that receipts were almost £12,000 down and costs £4,600 up. Bad weather and bus competition had hit them. Fuel and power cost more, and wage and other staff benefits had forced up expenditure. For every mile of route opened, the expenditure, said Acworth, was about five times what the LCC debited to its tramway account. The LCC could always charge part of street widenings for tramways to the street improvements fund, on which the ratepayers paid the interest. The LUT had no such opportunity.

The consolidation arrangements gave better hope. Of the preference holders 93 per cent and of the ordinary stock holders 97 per cent had exchanged their holdings for London & Suburban stock. The London and Suburban, said Acworth, would have a preponderant voice in LUT affairs, so that their own future meetings would be more or less formal.

In June 1913 Ealing showed itself more amenable than usual towards the LUT by agreeing to defer the purchase date by fourteen years, subject to the relaying of track and the substitution of side for centre standards. It subsequently considered whether to increase the wayleave amount and decided to defer the matter of the standards.

By this time the LUT was finding the purchase question a useful bargaining counter. For instance, it agreed in August 1913 to relay in Heston-Isleworth and realign in Twickenham Road if the council would extend the purchase date to 1930.

In August, trials for through running at Tooting were carried out, with LCC cars pushed across the gap between the tracks.

Evidence of the buccaneering tactics of bus operators appeared in the Report of a Select Committee on London Traffic which came out in August 1913. It stated that many people feared to ride in trams because of the way that buses ran alongside and prevented them from boarding or leaving the cars. It also showed that some bus fares were halved where they ran over tram routes.

One of the basic problems of the LUT was made clear at this time by A. J. Lawson in an article 'Greater London: its Area, Population and Traffic'. Lawson said that 635,228 people lived in an area served by the 55.34 route miles of the LUT. The large number of cars the LUT needed for its summer holiday traffic was an incubus. It had a fleet of 340. The

MET with 57 fewer cars could carry almost 27,500,000 more passengers in a year.

The use of trailers promised to produce economies. In its Bill in the 1914 session the LUT sought power to operate them and to permit overcrowding on cars on holidays, as well as agreement with local authorities to vary the periods and terms of compulsory purchase. Devonshire said that at times every available car was out but they could still not cope with the traffic. As the fleet was 250 per cent over normal needs, still more powered cars would be uneconomic. Trailers would provide the answer. Tooting was one of the LCC's chief routes for trailers and they had approached them about through running.

Ealing opposed trailers. Kingston thought them a danger. Wimbledon said they should run without the council's consent only at particular times.

The LUT gained a partial victory. An Act of July 31, 1914, gave permission to run trailers at rush hours – but at no time in Surbiton – if the existing service frequency were maintained. In bad weather the company could carry additional passengers not exceeding one-third of the number for which the vehicle was licensed. The Act also gave powers to enter into working and other agreements on through running.

At the annual general meeting on March 25, 1914, Acworth reported an increase of £5,662 in gross traffic receipts. In view of the competition it was satisfactory. When the LUT began operations, such competition could not have been foreseen. But they had touched rock bottom. There was still room for them, witness their 60,000,000 passengers in 1913 – more than in 1912.

The increased price of coal and other materials and exceptional expenditure on track and cars had pushed up costs by £12,000 but the special overhaul of cars was almost completed and the charge on the reserve fund for cast-welding joints was ended. There was £20,000 to place to renewals and £2,101 to carry forward. A 1 per cent dividend on preference stock for 1913 was recommended.

The tide seemed to be turning again but all hopes of a sustained improvement were dashed by the outbreak of war.

Two road works need to be chronicled at this stage. The administration of Kingston Bridge had been vested in Middlesex and Surrey County Councils in 1911 and the bridge was widened in 1914. In 1914 also the Great West Road Bill was passed. Its terms precluded the use of the road for public transport without the consent of Middlesex County

Council. It is a pity that the opportunity was not then taken to include a central reservation which might have given the LUT a Brentford avoiding line. Liverpool had shown the way with its new Bowring extension.

Early in 1914 Stanley had been knighted in recognition of his services to London transport.

Top: Type T No. 321 poses outside the 'Duke of Wellington', New Malden, in 1906. (*Courtesy Malden Library.*) *Centre:* Type T No. 321 negotiates the railway arch in Kingston Road, New Malden, preparatory to the Board of Trade inspection of Raynes Park extension in April 1907. (*Courtesy Malden Library.*) *Left:* Type W No. 243 at New Malden on Tooting–Hampton Court service.

Type W (later U) No. 287 eastbound in Coombe Lane, Raynes Park.

Type W (later U) No. 269 leaves a loop in Worple Road, Wimbledon, while a horse bus pulls over to allow passage.

No. 200 of type W inaugurates the Raynes Park–Wimbledon extension in May 1907. Note widening in progress in Worple Road, Wimbledon.

Top: Type W (later U) cars Nos. 272 and 274 stand alongside in Worple Road, Wimbledon. *Centre:* Type W No. 156 in Wimbledon Broadway on Summerstown–Wimbledon Hill service. *Bottom:* Type W No. 260 on Wimbledon Hill–Summerstown service rounds the corner at Merton (Grove). Site of future (1926) South Wimbledon tube station on extreme right.

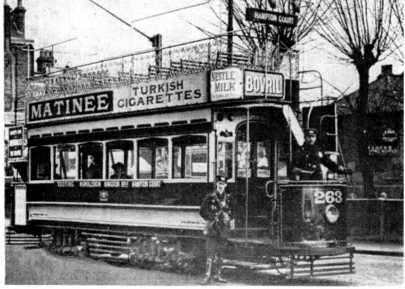

Top: Type W No. 180 on Summerstown service in Plough Lane. Future site of Wimbledon football ground on left. *Centre:* Wandle Bridge, Merton, about 1908 with Tooting-bound car. *Bottom:* In their military smartness the crew match the immaculate condition of type W No. 263 at Longley Road, Tooting, terminus. (*Courtesy R. B. Parr*)

Chapter 14

Wartime Readjustment

The war did not immediately hit transport undertakings. On the LUT half fares for children were made available from November 1914. They were valid between 8.30 a.m. and 5.30 p.m. on weekdays and up to 2 p.m. on Saturdays.

The Underground group assumed control of the London & Suburban and subsidiaries in May 1915, when the LUT offices changed to Electric Railway House, St James's Park, headquarters of the Underground. The BET retained a large holding in the London & Suburban until November 1928, when it sold its shares to the UERL.

The 1914-15 results made gloomy reading. Passengers were 1,211,342 down on 1913. Car miles were down by 410,335 to 8,996,891. But with average receipts per car mile no more than 8.47d, gross revenue was down so much that it hardly justified the ½ per cent dividend paid the preceding year on the 5 per cent cumulative preference shares. Hardly an auspicious heritage for the new management.

By this time 381 men, about 24 per cent of the staff, had joined the armed forces. Services were cut as reservists were called up. Later, as more men enlisted, more cuts were necessary. It was at least fortunate that so much track had been renewed before the war. The reconstruction had been completed at Hammersmith. Pending the sale of the Hammersmith lines to the LCC the LUT was paying interest and sinking fund charges on the sum advanced by the LCC for the relaying.

In October 1915 the Metropolitan Police sanctioned the employment of women conductors on London public service vehicles. The LUT took on its first 'clippies' in November. They received 4s 6d a day for a six-day week, plus a war bonus of 2s 6d.

One of the clippies was a Kingston girl who later became Mrs E. Seal. In reminiscing recently to Mr Bryan Woodriff she admitted that she enjoyed her work, though she found the LUT strict. The wages – more than £3 a week when she finished in 1919 – she considered good, as she had been earning only 15s a week as a dressmaker's assistant.

Mrs Seal and her colleagues wore a blue and white striped blouse, with a black tie and short jacket, and a long skirt. At first leather gaiters were issued. Later, as skirts rose to just below the knee, Mrs Seal wore boots costing about three guineas, which she had to buy.

She soon got used to jumping on and off a moving tram. Some conductresses found the tension of the trolley pole too much for them and had to ask the driver to 'run round' with it and rewire it at termini.

It was hard work on winter nights going up and down the unprotected staircases, especially if crowds from the Kingston Empire music hall were sitting on them.

It was a 'crime' not to collect fares and an ill-disposed inspector might well report one. An unpolished leather belt or a button undone might earn equal censure.

Mrs Seal generally kept to the Richmond Bridge–Ham Boundary route, though she recalled duty on the Tolworth route, where an old lady carried a footstool to help her to mount the car.

An early shift might run from 4.37 to 9.33 a.m., then from 10.56 a.m. to 2.13 p.m. A typical late shift might be 2.29–7.16 p.m., then 8.18 p.m.–12.20 a.m. There was a chance of a lift home at night in the staff car if one lived near the route.

Bad money from passengers had to be made up or passed on. Mrs Seal found that courting couples made the most unsuspecting receivers of 'dud' coins!

She could recall no staff canteens during her service. The wife of one of her drivers would meet the car near Fulwell and hand over a tea can. Mrs Seal would share some of this tea or take her own flask, with a tin of sandwiches to eat at the terminus.

One afternoon in 1919 Mrs Seal went to the dentist. Afterwards she reported that she felt too unwell to go on evening duty. While walking home in 'mufti' she was observed by an inspector. When she reported next day she was told that if she was well enough to be out she was well enough for work. She was sent to Chiswick where she was marched in between two other women and confronted by an official, who told her a dentist's note was not a doctor's certificate and therefore unacceptable.

The culprit was due to marry and gave in her notice to forestall disciplinary action!

In July 1915 a single-track connection of 159 yards was laid in between the MET and LUT at Acton for rolling stock interchange but was not

Wartime Readjustment

sanctioned for through public traffic. (Some MET cars were stored at Acton depot and later the LUT gained access to Hendon overhaul works.)

Reconstruction between Brentford and Hounslow, begun in 1914, was completed by early 1916. Even so, Brentford Council was still complaining to the Board of Trade and the Commissioner of Police about noise.

By the end of 1916 the company was getting short of rails for repairs and persuaded Hanwell Council to agree to the lifting of the unused track in Lower Boston Road.

Although gross receipts for 1916 were up by £2,190 to £336,335, the increased cost of labour and materials added £34,010 to expenditure and resulted in an adverse balance of £2,583. Traffic receipts were up by £1,503 as a result of the extra labour employed on war work in some of the districts served. Some 1,711,000 more passengers were carried in 1915–16 than in the preceding year and car miles were down by 1,250,000. Unfortunately the higher cost of materials meant that expenses were down by only £7,800.

At the annual general meeting of the London & Suburban in June 1916 it was said that even without the war it was doubtful whether the expectations of net revenue derivable from the LUT could have been realized. Upkeep of cars and track had cost more than expected and competition had not been countered to the extent hoped for.

A link with the old days snapped with the passing on November 22, 1915, of Sir George White, at the age of sixty-two. By that time the Bristol tramways had already established the bus feeder routes that would grow into a great undertaking serving large parts of Gloucestershire and Somerset and still retain much of its individuality under national ownership. By then also the brothers White had established the first aircraft manufactory in England, which in 1910 began to turn out biplanes and monoplanes, pioneer products of a now mighty organization which continues to bear the Bristol name.

In 1916 Sir Albert Stanley became Director General, Mechanical Transport, at the War Office. When Lloyd George formed his first cabinet, he became President of the Board of Trade and a Privy Councillor.

After two years of arbitration, litigation and negotiation, arrangements were settled at the end of 1917 for the LCC to buy the 5½ miles of line within the county in Hammersmith, and Chiswick power station, depot and offices, for £320,899, so realizing the County Hall aim that it should own or work all tramways in its area.

The deal, to come into effect with the end of the war, was an excellent one for the LUT, badly in need of such a windfall, and of scant value to the LCC.

The LUT had been able to insist that the power station, whose equipment was by then rather outmoded, be included in the purchase. The total sum included the £63,899 that the LCC had already advanced for track work in Hammersmith and £22,000 costs.

The LUT had arranged to take power for the whole system from Lots Road. It would have running powers over the acquired lines and the LCC would have powers over LUT lines in Middlesex, though the only access was from the peripheral Putney–Hammersmith–Harrow Road route.

As 1917 results were expected to be worse than those of 1916, and much needed to be spent on renewals, the application by a debenture shareholder for the appointment of a receiver was granted, pending a scheme to re-organize the capital.

The chairman of the committee of debenture holders said that it might be in their interest to seek some permanent arrangements with the Underground or its associates.

In November the company informed the councils in its area that it proposed to seek powers to be relieved of its fare restrictions and to abandon some sections.

The Hammersmith agreement was sealed in January 1918. The purchase price was fixed at £235,00, which the LCC was to pay on or before the expiry of a year after the end of the war.

In September 1917 a car descending Kingston Hill left the rails, crossed the road and knocked down a tree and a standard. It swung round on impact and the top was torn off and the body smashed. The only passenger, a woman, had head injuries. The conductress was badly hurt and detained in hospital, and the driver was injured.

On April 21, 1918, a car bound for Hammersmith left the rails in Brentford High Street, mounted the footpath, collided with a standard and fell on its side. As the day was wet, all passengers were inside. Eleven of them and the conductor were treated for injuries, and the street was blocked for some time.

Gross receipts for 1917 were up by £18,138, to £354,473, but expenses were also up, by £23,595. There was a balance of £60,028, which was needed to finance urgent renewals and repairs. Some savings were made from May 12th by putting back the start of Sunday services to about 2 p.m.

Wartime Readjustment

The Bill for a drastic reduction of the LUT's inflated capital, an increase in fares and the postponement of purchase dates by local authorities was passed on November 21, 1918.

The LUT's maximum fares, fixed by statute, had averaged about ½d a mile but in some places were 1d for three miles. This was satisfactory when only the best-paying routes were open and maintenance costs were low. From 8 per cent in 1903, the dividend dropped to 6 per cent in 1904, 3 per cent in 1905 and 1906 and 2 per cent in 1907. Nothing was paid in 1908 or thereafter.

Even the amounts set aside eventually became inadequate for repairs and renewals, and the other classes of shares reflected the trend. The preference holders managed to get 1 per cent in 1913 and ½ per cent in 1914, but there was nothing for them in 1915 and 1916.

The authorized capital of £2,500,000 was equally divided between cumulative preference and ordinary shares. Some £1,250,000 preference and £1,000,000 debentures had been issued.

As loans, including debentures, totalled £1,750,000, the total capital was £4,000,000, so that capitalization was as much as £40,000 per single-track mile. As much of the system had light traffic for some periods, the low fares made profitable operation impossible in the changed conditions.

Under a sensible rearrangement the arrears of dividend on the preference shares were cancelled. The company undertook to put its undertaking into a good condition as soon as possible. No dividend was to be paid until £400,000 had been spent on, or set aside for, reconstruction.

As from January 1, 1919, any revenue, after interest had been paid, was to be paid to a special reconstruction reserve for five years at a rate of £60,000 a year, for the next four years at £40,000 a year and for every subsequent year to January 1, 1928, at £30,000 a year.

Quite as important as the capital reorganization was the decision to appoint Christopher John Spencer to manage the LUT and MET. No better choice could have been made. Under the guidance of this extremely able Engineer-Manager the LUT was nursed back to something like health and a new phase of development ushered in.

Chapter 15

Spencer Makes His Mark

Spencer, whom we have seen advising Chiswick to try trolleybuses in 1910, was born in Halifax in December 1875. He began his career as a pupil on the Blackpool conduit tramway in 1889 and became electrician to the South Staffordshire Tramways in 1892, so that his experience of electric tramways actually antedated that of Robinson. His appointment as manager of the important Bradford Corporation Tramways in March 1898 was proof enough of his outstanding ability.

A thorough tramwayman, Spencer was early alive to the possibilities of trolleybuses. Early in 1910 he and members of Bradford Corporation Tramways Committee had travelled to see a Mercedes-Stoll line in Vienna and the Filovia system in operation near Lake Como. He had returned with enough faith to advocate a trial in Bradford.

Spencer, who took up his new duties on November 1, 1918, came south with a great reputation as the Manager of an up-to-date tramway system with useful trolleybus feeders. During the war he had performed valuable service for the Admiralty as an Assistant Director of its labour division. His move meant the end of his honorary secretaryship of the Municipal Tramways Association, of which he had been President in 1909–10, but in 1920 he became a member of the newly-formed Institute of Transport.

The *Tramway & Railway World* was right to predict a busy and interesting time for Spencer in his new sphere. His arrival coincided with the end of the war and the slow return to normality. It was a tough time for tramwaymen. There were about 5,330 miles of tram track in Britain, much of it defective because of lack of renewals in wartime. Little wonder that the new Woods-Gilbert Rail Remodelling Co. Ltd, specialists in renewing rails *in situ*, had full order books for a year.

The summer and autumn of 1918 saw a crop of mishaps. On August 16th a loose trolleyhead dislocated the service in Brentford by tearing down overhead on both lines. There was an accident between a tram and a lorry in Hammersmith in October and derailments occurred in Brentford in October and December. In November two well-laden cars collided in Cambridge Road, Kingston, at the convergence of single and double track. The fronts of both cars were wrecked. The drivers were uninjured but twelve passengers were hurt, three seriously.

The LUT was authorized to raise fares to a maximum of 1d a mile (1d for three miles for workmen) from December 9th. Routes were divided into half-mile stages. The date of purchase by local authorities in Middlesex and Surrey was deferred to 1950 or any subsequent period of seven years. As the next purchase date was 1924, a valuable breathing space had been secured.

The company was authorized to apply to the High Court to reduce its capital to £1,000,000 of mortgage debentures, £963,000 of 5 per cent preference shares and £660 of ordinary shares, a total of £2,623,000. For five years after the war the preference dividend was to be payable only out of revenue for the year concerned. The dividend on the mortgage debentures was limited to 4 per cent per annum.

In August 1919 Spencer was able to launch a big programme of track renewal. The first sections to be tackled were Hammersmith–Brentford and Acton–Hanwell. Also put in hand was the Hayes–Uxbridge section, for which £30,000 was authorized in July 1919.

Some 2,500 tons of rails were supplied from United States Steel Products Ltd at a cost of £40,000. The first 500 tons were laid on a three-quarter-mile section in Chiswick.

Spencer said that six British steelmakers had been asked to tender. Only one could do so. The Americans offered a fixed price and better quality steel. The cost of relaying was £30,000 a mile, compared with £12,000 before the war.

The work in Ealing included replacing centre standards by side poles and some realignment, all at a cost of about £40,000.

Gross receipts for 1918 were £362,427, an increase of £7,954, but expenses were up by the staggering total of £81,970 to £373,831, resulting in a deficit of £11,405. The total deficiency carried forward was £94,050, but was fortunately extinguished by the capital reorganization. Acworth said that the loss in 1918 had been £11,404. The wages bill was 60 per cent higher and that for materials was more than 100 per cent higher than before the war.

The year 1919 was not without mishaps. In May a car had its front wrecked and its staircase torn away when it collided with a steam wagon in Acton. Brentford remained a black spot until the track relaying caught up. Several cars were derailed and overturned and passengers were injured. Another incident, with a steam roller, took place there in October, after track work had begun. A tram was pushed into the excavation but fortunately remained upright. Three passengers were slightly injured in December when two cars collided in Surbiton.

With the complete changeover to Lots Road for bulk supply on October 31, 1919, Chiswick power station became a substation, incorporating some of the original switchgear.

At least the increased fares brought some comfort. Gross receipts for 1919 showed a healthy rise, up by £134,115 to £496,542. But the introduction of an eight-hour day, combined with labour awards and a continued rise in the cost of materials, increased expenditure by £122,855 to £496,687, so that there was a deficit of £71,577 to carry forward to 1920.

In February 1920 the LUT asked the Board of Trade to allow the maximum speed to be increased to 16 m.p.h. Some local authorities objected, among them Hammersmith, which wanted to impose a 12 m.p.h. limit.

The first order issued by the new Ministry of Transport under the Tramways (Temporary Increases of Charges) Act of 1920 authorized the LUT to increase its ordinary fares by 50 per cent. The new minimum was 1½d for two stages, 2d for three stages and 1d for each additional stage. The new fares added 1d to the 6d fares. By agreement the bus fares in the areas concerned were equated.

In the Thames Valley the LUT was facing a new competitor. Electrification of inner suburban lines, inaugurated in 1915–16, enabled the LSWR to hit back at the LUT, which had made such inroads ten years or so before. In 1913 LSWR steam trains were carrying 25,000,000 passengers over lines subsequently electrified. In 1919, after conversion of the Waterloo to Wimbledon via East Putney, Hampton Court, Shepperton, Kingston 'Roundabout' and Hounslow loop lines, the electrics were carrying 48,000,000.

Ashfield returned to the Underground Group in 1919 and was appointed Chairman and Managing Director, Devonshire resigning as managing director of the tramways on July 17th. In 1920 Ashfield was raised to the peerage as Baron Ashfield of Southwell.

Spencer was alive to the need for more economy in operation. The war had deferred the question of operating trailers. Spencer had other ideas. In the autumn of 1919 he and three other officers from the Underground

Group went at the behest of Ashfield to the United States to confer with tramway officers. Back home, Spencer wrote 'Some Notes on Transportation in America'. They show an original mind and a sense of humour. British trams, he said, still bore a remarkable resemblance to their horse predecessor. The driver still stood behind a steel dash and 'in his dreamy moments appears to see a pair of horses in front'. Apart from the use of the magnetic brake he still retarded the speed of his car by the strength of his right arm. Too much space was wasted by needlessly isolating the driver from the passengers.

The new and economical one-man cars he saw in the States impressed Spencer and he would apply their lesson.

The section between Kings Road junction and Ham Boundary, latterly a shuttle car working, lost its regular service in May 1920 but occasional journeys continued for some years to serve the Leyland (now Hawker) factory, situated a little beyond Ham terminus.

Gross receipts for 1920 were up by £36,200 to £532,742. As expenses were £482,597, there was a balance of £50,144. The improved finances were matched by an improvement in services with effect from March 30, 1921, when the interval was cut from four to three minutes between Shepherds Bush and Hanwell from 7 to 9.30 a.m. and from 4.45 to 7 p.m.

A slight adjustment in fares on the Tooting–Hampton Court route cut fares by 1d between principal points. Cheap returns were tried on this route and subsequently between Richmond Bridge and Kingston.

The better returns were maintained in 1921, with gross revenue of £584,030, a rise of £31,288. Working expenses were down by £9,024 to £473,573. The balance on working was £10,456, an increase of £60,312. There was a surplus of £59,474 as a reduction of the deficit of £70,882 brought forward.

Spencer's interest in one-man operation bore fruit in January 1922. He rebuilt MET No. 132, one of that company's 4-wheel single-deck cars supplied for the Alexandra Palace services, fitted it with a vestibule, the first in London. Transferred to the LUT as type S, No. 341, it was put into trial service on the Richmond Park Gates–Tolworth route. The cost of purchase and alteration was £1,168.

No. 341 was indeed a novelty. Passengers boarded and alighted at the front. An automatic ticket machine at each end enabled the motorman to issue up to five tickets by press button. The body was placed on an 8-foot 6-inch 4-wheel Brush truck with flexible axle suspension and 33-inch wheels. As altered the car seated 30 instead of its original 36 passengers.

London United Tramways

This car had a Spencer-Dawson hydraulic track brake, a device of Spencer's. Its handle was held normally in the 'on' position by a spring. In this position power was cut off from the controller. A light spring catch on the handle was held in place by the weight of the driver's hand. When pressure was removed, the handle was immediately released. The car also had rheostatic and hand brakes.

An ingenious 'road guide' worked off the axle showed passengers the position of the car and the route for half a mile ahead. Powerful headlights were mounted on the roof heads. Instructions on panels by the doors read 'Please Have Exact Fare Ready' and 'Please Pay as you Enter Car'.

An ornamental poster on a board running along the roof edge read 'Always be Careful' and was flanked by a Safety First triangle sign. Altogether the car brought a look of modernity to the route.

The through running at Tooting which Robinson had sought in vain to establish came about on May 2, 1922, though hardly in a manner he would have approved. The LUT handed over the Wimbledon Hill Road–Longley Road section to the LCC for operation. A change pit was installed at Longley Road and the LCC's services 2 and 4 from the Embankment were extended to Wimbledon, which then enjoyed modern cars of the LCC's 1777–1851 batch, built in 1921–2 and classified E/1.

The LUT Summerstown service became a shuttle to and from the Merton end of Haydons Road, though cars necessarily had to traverse Merton Road and Wimbledon Broadway to and from Fulwell. The Hampton Court route was cut back to the Wimbledon loop.

At the same time the LCC exercised its Middlesex lines running powers to the extent of extending its route 26 from Hammersmith to Kew Bridge and shedding some cars at Chiswick.

By this time the Richmond Bridge route was in great need of repair. Its traffic was light and hardly justified expensive relaying. Spencer considered that it might be suitable for trolleybuses.

At the Tramway Congress at Bournemouth that year delegates saw a demonstration of a new double-deck, front drive trolleybus built for Leeds Corporation by Trackless Cars Ltd, with bodywork by Blackburn Aeroplane & Motor Co. Ltd. The LUT borrowed the vehicle for trial at Fulwell.

About the same time a Middlesex County Council committee, fresh from a visit to study reserved track in Birmingham and Liverpool, proposed that the Hanwell Bridge–Southall section be reconstructed as sleeper track. In December the council approved the scheme in principle. The road was to be widened to 110 feet, with centre tramway reservation. The LUT was to pay the council £70 annually plus a lump sum representing the saving resulting from sleeper-tracking. The board approved but the project lapsed. Such a development, carried out at that time, might have altered the course of later LUT history.

Another weak link was removed when the Hounslow line was cut back from the 'Hussar' to the 'Bell' on Wednesday July 5th, 1922. Since the authorized Staines extension had lapsed the section had been of little value. The traffic was light and the track had deteriorated.

The Hanwell–Southall section was relaid *in situ* at a cost of £27,000. Other reconstructions approved at this time were: Twickenham (York Street)–Stanley Road (£32,700); Busch Corner–Twickenham (£60,700); and Southall–Hayes (Canal Bridge) (£19,400). Welding of joints between Merton and Kingston at a cost of £15,000 was also sanctioned.

In 1922 a Thomas W. Tinkler left the service of the LUT. The circumstance is interesting in that Mr Tinkler in 1956, when he was seventy-three, recalled some of his experiences. He told a reporter that in 1901 he had been a depot boy at Richmond. He had to trundle a ticket box in a barrow across old Kew Bridge and put it on an electric tram to Chiswick Depot. Tickets were issued daily and unused ones had to be returned each morning.

Tinkler had begun his career as points boy at Shepherds Bush at the age of fifteen. For five years he was a conductor on the Hammersmith–Hampton Court route. When Fred Karno opened the Karsino on Tagg's Island, Hampton, Tinkler left Chiswick on his tram at 6.30 a.m. He was still on it at 4.15 a.m. the next morning. This was exceptional but a 12–14 hour day was not.

Kempton Park race days, he recalled, were a harvest for conductors. People would walk from the course to Hampton Church. To be sure of getting a tram they would first catch one going the other way, to Hampton Court, and so be able to board a northbound car at the terminus. So full would the Hampton Court cars be on the short stretch from Hampton that the conductors could not take all the fares and would pocket the pennies instead. The management must for once have turned a blind eye!

The LUT, said Mr Tinkler, had merit and demerit marks. If a driver was 2 minutes late at a terminus, he earned a demerit mark. Sixty demerits meant the sack. Merit marks could be gained for aiding the company, as, for example, helping to remove a cart whose wheels had been wrenched off by being stuck in the rails.

When Mr Tinkler left the service he had 300 merits and had never been in an accident. His father retired after twenty-three years with a clean sheet – he had driven the first electric tram out of Chiswick depot for the initial run between Shepherds Bush and Kew Bridge, and the first electric between Shepherds Bush and Acton.

Mr F. E. Nutting, still employed at Fulwell, joined the LUT in 1923 as a conductor. In his first year he worked from May to the end of September and was then put off. He earned £3–£4 a week for two day shifts totalling nine hours. One week's paid holiday was granted. One of his earlier recollections is of weekend crowds joining LUT cars at Tooting terminus, bound for Hampton Court and laden with accordions and beer crates.

Mr Nutting also recalls Miss Janet Pearson of Kingston Hill who acted as a 'Lady Bountiful' to Fulwell-based crews. She dispensed hot coffee and sandwiches, which were collected by the crew of the first car to Kingston Hill, and she also gave annual garden parties for the men. The still-extant LUT Athletic Club at Fulwell competes in indoor games for an annual Janet Pearson cup.

Chapter 16

Mild Revival

By the middle of 1922 the Associated Equipment Co Ltd, then of Walthamstow, and an associate of the Underground group, produced a new 4-wheel single-deck trolleybus, the prototype of a design which it supplied subsequently to Shanghai and Singapore. It was first tested on the Mitcham–Croydon and Croydon–Penge/Crystal Palace routes of the SMET. By early February it was being put through its paces on the Merton–Summerstown route. On February 10, a cyclist collided with the unfamiliar vehicle in Haydons Road as it slowly moved away from a service test stop by Plough Lane corner. He was killed but the verdict was accidental death, AEC's driver being cleared of blame. The widow brought an unsuccessful action.

It seems likely that the vehicle used a skate trailing in the rail to provide the negative return and that trials on the LUT lasted for only a short time. The valuable researches of Mr J. C. Gillham, as summarized in *Modern Transport*, continue the story.

A second wire was erected along the route in April/May and turning circles put in at Summerstown and at the recreation ground near the Merton end of Haydons Road – though the road layout at this point does not appear to be very suitable for the purpose. New trials began in June but it is not clear what vehicle was used. It may have been from AEC or perhaps a borrowing from Bradford or another operator. In July a second vehicle was on loan.

In September, says Mr Gillham, the recreation ground circle was removed – possibly for reasons of traffic interference – and the double wiring from Summerstown cut back to the fairly wide T junction of Haydons Road and Queens Road, on the south side of Haydons Road railway bridge.

By the end of June 1923 double wires had been restrung along much of Haydons Road and in August they were extended up to the Haydons Road/Merton High Street junction. Mr Gillham records that in August and September a vehicle was on regular test but only between Summerstown and Queens Road, where it was manhandled round.

Children who assisted in the process were rewarded with a free ride. The double wires along the rest of the route to Merton were apparently used only when the vehicle was on its way to or from Fulwell depot. The negative wires were removed some time in 1924.

It is curious that so little information has survived about the trials, certainly the most extensive in London since the LCC trials between Woolwich and Eltham in 1912. It would be valuable to have a record of the vehicle and know more about the turning circles.

The experiment was obviously conclusive enough to prompt the LUT to go ahead with a full-scale conversion. The Richmond Bridge section was selected. The choice was sound. The traffic was light, the track in a bad state and the area close to a depot.

On November 14, 1923, a Bill was deposited for the complete conversion for trolleybus working of the Richmond Bridge–King Street, Twickenham, section and double-wiring thence to Fulwell, with a layout to allow the depot to be entered from both Wellington and Stanley Roads.

To provide a turning loop as close as possible to the bridge the LUT proposed to use Cambridge and Clevedon roads. This part of the plan aroused opposition, not unjustified as both roads are, or were, quiet side roads and are narrow. The junction of Cambridge Road with Richmond Road would have been awkward to negotiate and cause more traffic congestion then the stub tram terminus in the main road.

Both Twickenham council and Middlesex County Council opposed the Bill. Twickenham Highways Committee suggested that Cambridge Park, Alexandra Road and Morley Road would form a more acceptable loop. But these roads are only slightly more suitable and are situated inconveniently far back from the bridge foot.

The scheme had a poor local press. The *Richmond & Twickenham Times* took a rather pro-motorbus line. In its February 9, 1924 issue readers learned that '. . . the trolleybus has proved of great use in big centres of population but we do not know of a single town where it is run in competition with an up-to-date fleet of motor omnibuses.' A leader entitled 'New Tram Terror' in the March 1st issue warned residents of East Twickenham against any 'softness' by the highway committee in regard to the use of quiet roads by the LUT.

At the ordinary general meeting of the London & Suburban Traction Company on March 5, 1924, Lord Ashfield cautiously said that in some

Mild Revival

instances where track needed renewal they had thought it expedient to resort instead to trackless trolleys. The field for tramways development was limited, and the 'device of trackless trolley vehicles' would not, he imagined, widen it.

In view of the opposition in Twickenham the LUT decided to withdraw the Bill, seek powers to abandon the section and hand over the route to the General for motorbus operation.

Given that the tram route had to go, the public gained the advantage of a new through link with Richmond and beyond. But it is a pity that the LUT's first serious essay in trolleybus operation should have come to naught for lack of a good turning loop. Otherwise, one or two more weak tram routes might have been converted to trolleybus over the next two or three years and the company enabled to concentrate its tramway activities on the better-patronized, better-kept sections. One could further speculate that such a policy might have justified the modernisation of the whole of a reduced tram fleet, by, say, 1926, with incalculable results.

At the same London & Suburban meeting Ashfield had to report a setback in results in 1923, after results which had marked progress in 1921 and 1922. The friendly restraint which had held off the General buses had left a field ready for 'pirates' to exploit. Now the General felt that it could no longer restrict its operations in tramway areas if others were allowed in.

The tramways, said Ashfield, were under a statutory obligation to issue cheap returns for workmen. They brought heavy but unremunerative traffic. Buses now seated up to fifty-four passengers and were 'more level' with trams, the advantages of whose special track tended to disappear.

But tramways should not be abandoned. They had £6,500,000 capital invested in all their tramways, which carried more than 1,000 million passengers in 1923. The tramways deserved support and protection and had a place in a complete co-ordinated system.

Net income was only enough to pay 2½ per cent on the cumulative preference shares. The arrears of the preference dividend would then be £345,000, or 21½ per cent on capital.

(Net revenue for 1922 was £126,798 and there was a surplus of £68,652, of which £11,408 was applied to work off the 1921 deficiency. The balance of £57,244 was a welcome contribution to the reserve for reconstruction and renewals. Traffic receipts for 1923 were £425,996 and total receipts £67,778. Since 1918 £353,533 had been spent on reconstruction. When £400,000 had been spent, it would be possible to pay a dividend.)

On February 18, 1923, Lady Robinson died, aged seventy, at Bladon Lodge, South Bolton Gardens, where she had been living quietly for some years.

On August 16, 1923, a tram collided with a bus in Chiswick High Road, wrecking the upper deck of the bus. A number of passengers were injured and two were later reported to have died.

The Richmond Bridge–Twickenham section closed on October 1, 1924. The *Richmond & Twickenham Times* of October 4th had an editorial 'The Last Voyage (by our own Tram Mariner)'. It ran: 'We put to sea from the harbour at Twickenham – whew! it was a rough night! – under the skilful guidance of Driver S. Cook, who has been with the old vessels for many years, and was accompanied by that sturdy helmsman Conductor Wright, who had proved himself, man and boy, a right good conductor, since the ships of the road first put to sea. . . .'

There were few aboard. They included two lovers, a medley of raindrop dodgers and the reporter. 'It seemed as if the wraith of Sir Clifton Robinson was standing in the doorway of this, the last tram, and reminding us that the company had since its inception earned the gratitude of the travelling public, especially for the courtesy of their servants, on the very heavily used Hampton Court lines.'

The writer wound up this not unkindly notice by saying that there was reason for congratulation that the service was not otherwise to be curtailed and the the board was still devising better, quicker and cheaper travel.

The one-man car had proved a success in Kingston and also on the Boston Road route to which it had been transferred in 1924, but being sandwiched between double-deck cars could not show its capacity as a timesaver. Spencer therefore decided to extend the experiment.

The car was rebuilt at Hendon as a six-windowed saloon and mounted on Brill bogies. It was joined by three other conversions, two of them of Nos 175 and 275, the single-deck 'social saloons', which had been little used for some years, and the third of double-deck open-top car No 178, cut down and similarly rebuilt for one-man, PAYE, operation. Renumbered 342, 343 and 344 they joined No 341 on the Boston Road route, the four cars operating it entirely.

Nos 342–4 included an ingenious mechanical device by L. B. Hewitt, a technical assistant. The pneumatic doors worked in conjunction with

Top left: 'Electric Cars Stop Here If Required' notice. *Top right:* 'Electric Cars Stop Here' notice. *Above:* 'Cars Stop Here on Sundays During Hours of Divine Service' notice. *Below:* Parade of cars at Fulwell depot.

Top: LUT and LCC almost meet at Longley Road, Tooting, where type U (ex-W) No. 296 and type W No. 168 are in foreground, with LCC cars behind. (*London Transport*). *Centre:* LCC car at Wimbledon Hill terminus soon after the LCC takeover of the Tooting–Wimbledon service in 1922. *Bottom:* LUT car on Shepherds Bush service and (behind) LCC car on Hop Exchange service at Kew Bridge after extension of LCC route 26 over LUT in 1922.

Top: Contretemps in Kingston, probably in the early 1920s. (*London Transport.*) *Centre:* Type U (ex W) No. 300 in Clarence Street, Kingston, in 1922, on short-lived extension of Hammersmith–Hampton Court route. *Bottom:* Type W No. 240, later remotored, at Tolworth on Richmond Park Gates service in the early 1920s.

Top left: Mrs. E. Seal of Kingston was an LUT conductress in the 1914–18 war. (*Courtesy B. Woodriff.*) *Top right:* Type Y (ex Z) No. 48 in Haydons Road on Merton-Summerstown shuttle in the 1920s. *Centre:* Type Z No. 52 ran with this Barber six-wheel truck in 1909. (*Courtesy R. B. Parr*). *Bottom:* Cars Nos. 141 and 142 were converted to 'flood cars' in the 1920s to maintain a 'ferry' service on the Hampton Court–Wimbledon route at West Barnes, where the track was liable to be flooded. The traction gear was re-mounted high inside the body of one car and removed from the other car, which worked as a trailer. (*Fox Photos*)

Mild Revival

the steps. Passengers boarded at the front and alighted at the rear. On leaving they had to stand on a treadplate before the door would open. The door closed automatically when the passenger left the step. The airbrake handle had a deadman grip. Push bells replaced cords. The car seated thirty passengers on brown leather longitudinal seats.

On August 12, 1924, more than 200 wives and children of LUT staff took an outing from Hounslow to Folly Farm, Hadley Wood, riding in chartered LUT cars over the MET via Acton and Cricklewood to Barnet.

In September 1924 the Parliamentary Committee of Middlesex County Council reported on the council's Bill, which included a clause inserted originally to empower the council to provide a tramway reservation on Uxbridge Road. Advantage was to be taken to make this clause apply in general to all tram-served roads not less than 100 feet wide. Subject to Ministry of Transport consent, the council would have power to realign any light railway or tramway on a reservation.

The council expected the LUT to bear all the legal costs of inclusion of sleeper-tracking on the Uxbridge route in the Order which the Council intended to promote in connection with its projected Tottenham-Walthamstow light railway.

Net income for 1924 was £33,428. A deficit of £16,356 was carried forward.

By mid-1925 Spencer had carried out a notable modernization of cars. The T Class 1906-built canopied cars had been fitted with new lightweight Metrovick motors weighing 39 pounds per h.p. compared with the 81 pounds of the original motor. The cars could now accelerate from 0 to 12 m.p.h. in 8 instead of 9 seconds and to 20 m.p.h. in 24 seconds instead of 52 seconds. On a service test of six 5-second stops per mile they could average $13\frac{1}{2}$ m.p.h. compared with 11 m.p.h. The new Maag gears made them quieter running.

Some cars were redecorated in harmonizing colour schemes, such as light oak and white enamel, white and French grey, and cream and brown. Transverse seats covered with blue and fawn moquette were fitted in both upper and lower saloons, at a sacrifice of six places. For the first time a London road passenger vehicle had spring upholstered seating. New headlights were also fitted.

The remodelled cars were a credit to the company. The *Tramway & Railway World* wrote with justice: 'Under the deluge of criticisms levelled against tramcars, the public have almost lost sight of the fact that the modern tramcar, at any rate so far as London is concerned, is a very different vehicle from its pre-war prototype.' Unfortunately, because of the heavy capital tied up in tramways and the leeway in repairs to be made up after the war, their modernization in both London and the rest of the country was not as widespread as it could have been. It came late, too, when the modernized bus, profiting by the rapid development of the internal combustion engine during and after the war, and the improving trolleybus seemed to offer at least some of the advantages of the tram, without its costly track, on routes with light or medium traffic.

Bus competition was still a great menace. In March 1925 the situation was bad enough for the LUT to tell Twickenham Council that it could not undertake to relay track in London Road unless it were relieved of some of the bus competition.

Some relief came on the Uxbridge Road when the Ministry of Transport issued an Order limiting the number of bus journeys, to relieve congestion and protect the LUT from excessive competition. In September the LUT sought the view of passengers on the modernized cars.

Sir William Acworth died on April 13, 1925.

LCC 'invasion' of LUT territory was carried a stage further on Saturday May 22, 1926, when through cars began to run between the Embankment and Hampton Court via Tooting on summer Saturday afternoons, Sundays and Bank holidays. The weekend advent of the LCC E class was hailed in such places as Malden and Kingston. With their more rounded aspect and transverse seats in the lower saloon they invited invidious comparison with the now ageing LUT cars.

At least the LCC came by consent, and operated trams. There had been bus competition at the weekends since 1913 when the National Steam Car Co. Ltd began a Sunday service between Peckham and Hampton via Wimbledon. Steam gave way to LGOC petrol buses after the war, until 1923.

At the Tramways & Light Railways Association Conference in Torquay in June 1926 Spencer ably defended trams against those who accused them of obstruction.

Spencer said he had been a motorist for twenty-five years. He quoted an instance on the Great West Road when he had been unable to pass a slow-moving van. If the van had been a tram he could have overtaken without trouble. Trams, he added, were the most economical users of road space.

Traffic receipts for 1925 were £374,284 and expenses £360,551. There remained a deficit of £29,149 to carry forward with the previous deficit of £16,356.

Mild Revival

Unfortunately heavy interest and other charges upset the better traffic figures in 1926. Traffic receipts were up to £390,894 and expenditure only slightly up, at £363,698. Final income was £34,202 but a final deficit of £20,070 resulted, making a total deficit of £65,575 to carry forward.

The story was repeated in 1927, when, despite traffic receipts of £412,213 and expenditure of £367,786, a cumulative deficiency of £68,020 had to go forward. Stabilization of bus working over tram routes, combined with the improved cars and services, had an effect by 1928, when net income was up by £16,698 to £68,483, enabling the adverse balance to be cut to £53,431.

The improving trend seemed to justify the policy of partial modernization. Unfortunately the LUT had a good deal of lean for a little fat and it was apparent that some decision would need to be taken on the Thames Valley and Surrey lines, with their sporadic traffic, deteriorating cars and, in places, poor track.

Meanwhile a useful new facility was provided in inner West London from Wednesday November 28, 1928 when the LUT and LCC began to operate jointly between Acton and Putney, an extension of route 89.

Chapter 17

Modernization

Coming from the General Manager of all the Underground Group's tramways and the Chairman of the Tramways & Light Railways Association, Spencer's evidence before the Royal Commission on Transport in 1929 carried much weight. He rightly urged that competing bus operators should be subject to some of the labour, fare and stopping-place conditions imposed on tramways. Tramway companies should not have capital wasted by unrestricted competition. The tramways' burden of road maintenance and local taxation should also be lightened.

He gave telling figures to show the beneficial results of the London Traffic Act on the Uxbridge Road traffic:

year	revenue total £	per car mile	car miles	competing bus miles	tramcar passrs
1922	169,642	23.51d	1,731,543	780,000	18,250,156
1925	98,078	12.05d	1,952,904	3,250,000	13,149,895
1928	146,700	16.88d	2,086,200	2,350,000	19,700,000

In the summer and autumn of 1929 the Underground reappraised its tramways. The satisfactory performance of new and modernized cars on the MET – by far the most profitable of its three systems, with a generally balanced and good traffic and most track in good condition – led to an announcement that if Middlesex Council would renew the lease, the MET would be made 'one of the most efficient and comfortable tramway systems in Europe'.

Something of this optimism spread to the LUT. In December it was given out that the Group was considering the conversion of part of the Hanwell–Uxbridge section to a high-speed light railway and had applied to Middlesex Council for permission to lay a fenced-off roadside reservation. Spencer had hinted at such a possibility when he observed to the

Modernization

Royal Commission on Transport that reserved-track tramways were ideal for moving large crowds.

But the future of the Thames Valley and Surrey lines posed a problem. They formed almost a distinct system geographically and in their traffic. Indeed, with the market town of Kingston as their centre they were akin to one of the medium-sized provincial systems which were becoming increasingly vulnerable to the challenge of the motorbus and trolleybus.

Spencer was well qualified to appreciate the rapid strides made in trolleybus design in the 1920s. Fresh in the minds of some provincial managers who were faced with sporadic traffic and deteriorating cars and track, was the recent successful changeover at Hastings. The LUT had been unlucky in earlier attempts to run trolleybuses. Perhaps a fresh attempt would have a better chance.

A Bill was drafted for the 1930 Session seeking powers to convert all or any part of the LUT to trolleybus operation and specifically to substitute trolleybuses for trams on some twenty miles of route in the districts of Twickenham, Teddington, Hampton, Kingston, Surbiton, Malden, Merton & Morden and Wimbledon, and equip any additional routes authorized by the Ministry of Transport. It also contained powers to increase outgoings properly chargeable to revenue and so reduce the sum which the company had to pay in any year into the special reserve, as enjoined in the 1918 Act.

It was made clear that there was no intention, in the first instance, to do more than experiment in certain areas, of which the first would probably be Kingston. But general powers had been sought to permit trials wherever it seemed desirable.

Meanwhile tram services were improved wherever possible. An innovation was a new weekday route between Hammersmith and Brentford via Hanwell. The opening of the Wimbledon Stadium at Summerstown in 1928 gave a fillip to evening business on the Merton-Summerstown route. Extra cars were sent from Fulwell in the evenings.

Of the London tramway undertakings, the LCC alone had been in a position to make any additions to its mileage since the war. Admittedly many of its more ambitious plans had been turned down, including lines in central London to link some of the stub termini. New LCC routes served large housing estates in the south-east, though, surprisingly, they were not laid on reserved track. Now the LCC planned to extend to serve other new estates, at St Helier, Surrey, and Becontree, Essex, the latter to be reached over reserved track from Barking. In addition it proposed to enlarge the only central London link, the Kingsway Subway, to take double-deckers, and spend £5,000 on an experimental new car, which might be the first of a new fleet.

These projects, coupled with the knowledge that the Underground was perfecting new tram designs, seemed to add up to something of a tramway revival. Lord Ashfield told the American Electric Railway Association in November 1929 that the problem was to plan so that underground railways, tramways and buses might be used to best advantage in providing a unified transport system.

At the meeting of the Underground Group in February 1930 Lord Ashfield said: 'It has been the habit in recent years to depreciate the value of the services rendered by tramways . . . but we found that if steps are taken to bring the standard of comfort of the tramcars up to that customary upon the railways and omnibuses, and that if steps are taken to increase the general speed of operation of the tramways, the passengers carried by the tramways will increase in numbers, proving that tramways serve to deal effectively and economically with considerable volumes of traffic upon streets.'

The average tramcar speed was now 10 m.p.h., not unsatisfactory in view of growing congestion. Since 1925 a slow if cautious advance had been registered in the traffic receipts of the associated three tramways, which aggregated £1,275,552 in the preceding year, a 4 per cent increase on 1928. But they were still dogged by the impoverishment which beset the companies during the war and the subsequent depression.

'We should like to spend a great deal of money on our tramways,' continued Lord Ashfield. There was a particular need to spend much on the reconstruction and re-equipment of the Uxbridge Road line and to provide new rolling stock for trunk services. With regard to the replacement of trams by trolleybuses on the lighter-traffic routes, they would have to await the outcome of their Bill. There were difficulties but they hoped with the aid of holding or finance companies to do something to restore tramways to their original position.

It was still impossible for the MET and LUT to pay a dividend. They might be able to get further assistance from the Government under the Development (Loan Guarantee and Grants) Act of the last year towards meeting interest charges on further expenditure to carry out new works. In

Modernization

expectation of such assistance they had taken steps to order new tramcars.

Ashfield said in March 1930 that the 46 new cars which would be the LUT's share of 100 ordered by the Group would bring a substantially increased traffic to the Uxbridge route. (The other 54 were destined for the MET, whose future seemed assured when the Middlesex Council agreed to extend the lease by forty-two years.)

The 1929 results bore promise. Gross receipts were £469,194, a rise of £15,747. Net receipts rose by £5,006 to £65,997, making total income £73,573 and giving a balance of £17,389. The deficit of £36,047 was carried forward.

On May 16, 1930, a Lords Select Committee presided over by Lord Weston Wemyss approved the LUT's trolleybus Bill. Mr Craig Henderson, KC, for the company, said that no dividend had been paid on the ordinary stock since 1907 and none on the preference shares since 1914. Wartime conditions followed by intensive bus competition had seriously affected finances. Revenue in 1925 had dropped to £381,000. With improved rolling stock and services it had risen to £477,000, but it was still not enough to give a return on capital and allow further expenditure on track renewal. Road and track maintenance since 1918 had swallowed up £878,000 and the company could not pay into a reserve fund. The estimated cost of conversion was £248,000 of which the removal of track accounted for £80,000.

Local authorities had suggested petrol buses as replacement for trams in the area in question but the company considered trolleybuses preferable because of their greater acceleration and quieter running.

Spencer said that they would first equip the Twickenham-Kingston-Tolworth section and again stressed that it did not follow that they were going to supplant trams entirely.

The Act, which received the Royal Assent on August 1, 1930, set the limits of the conversion area at a point 350 yards north of Twickenham Railway Bridge and at Wimbledon Town Hall.

Little time was lost in starting work. By October trolleybus wiring had been installed by Clough, Smith & Co. between Twickenham and Teddington. On October 1st a new AEC-English Electric trolleybus was tried out at Fulwell. Save for minor differences in design it was similar to the 60 which the LUT had on order. Chassis and body were of standard AEC 6-wheel design. The lower saloon seated 16 in transverse and 11 facing seats. The upper saloon had 33 seats. The bus was powered by an 80 h.p. motor.

London United Tramways

Nottingham Corporation trolleybus No. 27, a Karrier-Clough vehicle, was also tried.

An important London tramway appointment of and one fateful for London Transport was that of Theodore (later Sir Theodore) Eastaway Thomas to be General Manager of the LCC tramways. Thomas was a former Underground Group officer who had entered the LUT drawing office in 1899 and by 1901 was engaged on the electrification of the system. When the Surrey lines were completed he became resident engineer.

When the LUT entered the Underground fold in 1910, Thomas went to Electric Railway House to undertake publicity and traffic work. In 1917 he transferred to the LCC as Tramways Superintendent.

Thomas, who had seen the LUT's electric trams enter service, would see them out, some thirty-five years later.

Net income for 1930 was £58,427, a drop of £757. There was a balance of £4,092 to be applied to reducing the deficit to £31,995.

Curiously, in view of the earlier experiments, the Summerstown route was excluded from the conversion scheme. On April 16, 1931, its working was taken over by the LCC which extended its route 14 (Hop Exchange-Wandsworth) to Wimbledon Town Hall via Summerstown, where a change-pit was installed.

The 100 new cars for the MET and LUT were to a design prepared by the MET's Manor House office. The production order was announced in August, 1930. Delivery was to begin in December 1930.

The design was the final outcome of the experience gained with the construction and operation of the five experimental MET cars, and embodied some alterations. For example, instead of equal-sized wheels, the cars were given maximum traction swing-bolster trucks, designed by the Underground and built by EMB; they had 28-inch driving and 22-inch pony wheels. (Car No. 396 received a different type of truck.)

The axle boxes had SKF roller bearings. In addition to a handbrake and magnetic track brake there was the British Air Brake company's straight air brake for service stops. LUT cars had two 70 h.p. GEC motors, in contrast to the BTH motors of the MET.

Construction was entrusted to the Union Construction & Finance Co. Ltd, Feltham. This company had a curious history. It had been registered on October 16, 1901, by Yerkes, who became Chairman.

The company lay dormant until 1925, when the Underground awakened it to undertake the modernization of Central London Railway rolling

Modernization

stock. It assumed its final name in February 1929, when the articles of association were altered. Two months later it completed the first of its experimental cars for the MET. It was as a result of trial operation with these cars that series production was put in hand of what became known as the UCC or Feltham cars.

Chapter 18

Finale

The UCC cars were designed to maintain the excellent average speed, stops included, of 12 m.p.h. travel at more than 20 m.p.h. on favourable stretches and accelerate on straight track from rest to 20 m.p.h. in 20 seconds.

The body shells and underframes were jig-built by UCC and delivered to Fulwell for the trucks to be fitted. The estimated cost of each car was £3,420.

Most passengers complied with the request to alight by the front exit, though this was not compulsory. The car could not start until all doors were closed. The exit arrangements saved time once passengers had become used to them.

When the brakes were applied, a 'Caution' light showed on the dash. Around it were the words 'Passengers Alight Front and Rear End'. Later, a rubber 'Stop' flag which shot out when the door opened was fitted.

Entry was by a pair of double doors 2 feet 3 inches wide.

A handle on each platform enabled the conductor or a passenger to operate the air brake in an emergency. The LUT drivers preferred the air brake to the magnetic. There were two sanding systems.

The route number appeared on each side of the blind destination box.

Of all the LUT routes the Shepherds Bush–Uxbridge had the most consistent traffic, with a useful midday peak of shoppers and passengers riding between home and work. It was also free from very sharp curves. On most other routes the clearances were too tight to allow Felthams to be used regularly – though it is believed that a 'Baby Feltham' design was also considered. Even when leaving Fulwell the first Feltham is said to have struck a pillar, and there was an awkward dip by Brentford gas works that would have been a hazard in regular operation there.

On Christmas Day 1930 an LUT and a MET Feltham were tried out in

Finale

Southbury Road, Enfield, which was closed for the purpose. Other tests were made at night between Kingston and Hampton Court.

The Felthams were at once popular. The *West London Observer* in a leader 'New Luxury Tramcars' in its January 9, 1931 issue said: 'The first eight have now been delivered . . . they attracted considerable attention. . . . In speed, therefore, as well as comfort, the new tramcars will bear comparison with any other form of transport.' It was a long time since the London lay press had written so kindly of trams.

Clough Smith was making good progress in installing some four miles of double overhead between Twickenham and Kingston Bridge. Of two buses delivered, one was in daily use for training motormen as drivers.

The order to build 60 trolleybuses had also gone to UCC. The chassis were of the AEC Renown 6-wheel type used for the LT type buses of the LGOC. Thirty-five had English Electric 80 h.p. motors and 25 BTH 82 h.p. motors. There were pedal-operated controllers and Dewandre vacuum-servo brakes.

Externally the bodies were similar to those of the LT enclosed-staircase type bus. They seated 24 on the lower and 32 on the upper deck, all but No. 60, which had curved instead of straight stairs and held 27 down and 29 up. The buses could reach 29.4 m.p.h. on the level and accelerate from 0 to 20 m.p.h. in 9.2 seconds.

No. 60 was different also in having seven small bumpers, mounted one above the other on the offside rear.

The livery was similar to that of the Feltham trams. The interior was of silver-grey woodwork. In the lower saloon there were panels of blue rexine on plywood. The upper deck was mainly in red. Downstairs the seats were upholstered in grey moquette with a green lozenge. Upstairs they were mainly red.

The outer sheeting was aluminium. Bearers were teak, except over the rear axles, where aluminium was used. Frames were ash and teak, with aluminium alloy gussets.

The wheelbase was 16 feet 6 inches. The body was 27 feet $2\frac{1}{2}$ inches long, 7 feet 5 inches wide and 14 feet $3\frac{3}{4}$ inches high.

Except for the trolley booms, the vehicle looked much like a bus, as the motor was placed under a bonnet. Unusual features were the central front upper deck window, at first arranged to swivel about a horizontal axis. Later fitted was a large central headlight.

Trolleybus service between King Street, Twickenham, and the Savoy

Cinema, Teddington, where the roads were wide enough for turning circles, was officially inaugurated on Saturday May 16, 1931. A temporary tram crossover was put in outside the church of SS Peter and Paul, Teddington, but was needed for only another month.

The Mayor of Twickenham (Alderman J. Owen), started the inaugural bus, No. 4. Both Nos 4 and 5 were decorated and carried the Mayor, aldermen and councillors, other guests and LUT officials to Fulwell for a celebration lunch.

Presiding at lunch was Frank Pick, the Managing Director of the Underground group since 1928. He playfully blamed busy roads for his inability to join in the inaugural ride and then hinted at the impending unification of London passenger transport. Amid laughter he said: 'These vehicles are to be taken over by a board set up by the government, but we have this satisfaction: we have not paid for them yet. The board will inherit our debts as well as our trackless trolleys.'

Frank Pick, born in 1878, had joined the North Eastern Railway in 1902. He went to London in 1906 when Sir George Gibb became Manager of the District and London Electric railways. He joined Stanley's staff in 1907, and became traffic development officer in 1909 and Commercial Manager in 1912.

It is pleasant to record that the Mayor of Twickenham considered that there was a future for trams. Certainly there was no question of the service that they were performing, for Pick told his audience that their trams were carrying twice as many passengers as in 1902. There was no need for pedestrians.

In describing the inauguration the *Tramway & Railway World* measured its words well. On the lines scheduled for conversion, it stated, much of the track would soon need costly renewal. The traffic was not very dense and trolleybuses would handle it more economically. The fact of the Feltham cars showed that the Underground would retain tramways where conditions favoured.

The fallacies of anti-tram factions were exposed by Spencer at the Tramways, Light Railways & Transport Association Conference at Margate that June. Mr E. H. Edwards, Chairman of the Association and Managing Director of the South Lancashire Transport Company, who had been Robinson's resident engineer when Fulwell depot was built, said he had recently ridden on the LUT in the only really modern tramcar he had been in. He had been told that the track was twenty years old, but there was no noise at all. The car was very fast, with marvellous accelera-

Finale

tion. He considered the latest LGOC bus a poor third in comfort or silence after the new LUT trolleybus and the Feltham tram.

The eight vehicles delivered enabled public service to begin at once between Twickenham and Teddington. By June 5th the eastward spread of conversion had progressed enough for trial trips to be made in Kingston. History was repeated when the Mayor of Kingston (Councillor W. Bell), Spencer and others made the first official trolleybus ride across Kingston Bridge that morning.

On Monday June 15th the Teddington service was extended to the new Park Road loop in Norbiton. The LUT had preferred to connect the Richmond Park Gates and Kingston Hill routes via Queens Road. After much debate Kingston Corporation by a majority of one opted for a link by the parallel Park Road, slightly nearer Kingston. The Ministry of Transport approved the choice and the LUT acquiesced, losing no time in erecting poles and wiring. The end of the two tram routes were abandoned. Trams from Tolworth and the Dittons were cut short at Eden Street.

Of the conversion an LUT official said, rather inappropriately, 'We are going full steam ahead with the construction of the vehicles, the training of the drivers and the fitting up of the routes.'

Between Wednesday July 15th and July 29th, when the Eden Street–Tolworth conversion was inaugurated, a shuttle tram service linked Surbiton and the Dittons until the Dittons–Kingston Hill loop route was inaugurated. For a time there was a Dittons–Tolworth via Kingston Hill loop service.

On August 15th Clough Smith completed the overhead work between Norbiton and Wimbledon, after just under ten weeks' work. By the end of the month a joint tram/trolleybus service was operating between Hampton Court and Wimbledon.

The advent of the trolleybuses was the signal for invidious comparison in the press. But not all comment was wholly unkind. The *Wimbledon Borough News* struck a wistful note in 'Old Friends', an editorial in its August 21, 1931 issue: '. . . a sad sight to see old familiar objects disappear, and it is quite possible that many people, whilst riding in comfort in the new trackless buses, will give a tender thought to the old trams which, for about a quarter of a century, rattled and swayed along the Worple Road. After all, they have been very useful . . . and to many of us they will always be associated with a number of happy incidents'.

In the next issue a sharper note was struck. '. . . those rare old antiques, the LUT tramcars on the Wimbledon–Hampton Court route, are rattling their last. They may be swaying and nearly hurling the passengers down the precarious staircases more violently than ever, but these are only the effects of the dying gasp. . . . Already the stopping places bear the indication, "Trolley 'Bus Stop by Request'". The same journal, more amusingly, likened the trams' reversed staircases to the approach to the fighting top of a man-o'-war!

The Wimbledon–Hampton Court trolleybus service began in full on Wednesday September 2, 1931. At first the buses used the St George's Road/Francis Grove loop at Wimbledon, though in the opposite direction to the trams, but on December 15, 1932, they were extended to a new turning circle outside the new town hall, just to the south of the railway station. At the same time the LCC tram service was cut back to the same point.

The through LCC weekend running to Hampton Court necessarily ceased on August 31, 1931.

The trolleybuses caused unexpected interference with the London National and Regional broadcasts of the BBC in the areas served. The trouble was cured when the LUT ordered 120 choke coils, each weighing 40 pounds, which were fitted to the roof behind the base of the trolley arms.

This step ended one of the few complaints from the public, though in September 1931 the *Surrey Comet* carried a letter complaining of the screech of the trolleybus brakes!

The greater part of Fulwell depot was turned over to trolleybuses and the remaining trams were restricted to five tracks. The five tracks on the north side and the three-track section on the south were adapted for trolleybus overhaul. Tramcar repairs were transferred to the Hendon depot of the MET.

Spencer ably summarized the case for the trolleybus conversion in a paper published in June 1931. The alternatives, he said, were relaying plus more Feltham cars, or abandonment in favour of petrol buses. It would have cost the company £550,000 to rehabilitate the tramways in the area concerned. It cost only £230,000 to convert to trolleybus working, including the £66,000 paid to road authorities in respect of abandoned tracks.

The company considered the area its most favourable one for vehicles of lower capacity than trams. Its peak/slack traffic ratio was relatively good and the traffic was nowhere too dense for rubber-tyred vehicles.

Finale

The trolleybuses were costing 13 per cent less per car mile than the trams and earning 26 per cent more revenue, even though as smaller vehicles they ran about 24 per cent more car miles.

The next part of the paper presaged, probably unintentionally, the wholesale tramway conversion that the future London Passenger Transport Board would usher in. 'Where traffic is dense and peak demand heavy we have not yet found a successful method of replacing tramcars,' he admitted, though they were trying to design a trolleybus of sufficient capacity to replace trams in some other districts. Future policy turned on that.

Spencer recognized that the novelty element had brought new business and that it was invidious to compare new trolleybuses with thirty-year-old tramcars. It might have surprised him could he have known that the new Feltham trams would survive that long, albeit in new surroundings, and still not suffer by comparison with the buses that finally replaced them!

He gave the following figures for trolleybus operation between May 16 and December 31, 1931:

	£
Receipts	67,326
Operating costs	52,756
Net receipts	14,570
Miles run	1,048,839
Passengers carried	11,929,417
Passengers per car mile	11.30

The Tramways, Light Railways & Transport Association met in London from May 18 to 20, 1932. Spencer read a paper on 'Tramcars and Trolleybus' to members, at Fulwell. He said of the Feltham trams that they had attempted to secure the utmost seating and standing accommodation in minimum space, and smooth, silent running. The only waste of space, and that not extravagant, was the cabin. Competition, he alleged, was probably more severe in London than anywhere else in the world. But for such an effort to keep up to date, the tramways could probably not have subsisted.

Spencer referred to the general lack of a midday peak in London compared with provincial towns. London had to have high-capacity vehicles or the service became too extravagant. On the busiest London tramways the high rate of acceleration with the magnetic brake caused terrific rail wear but without such a rate the high speed of London trams could not be maintained. The LUT drivers preferred air brakes for general braking.

The visitors were conveyed by Feltham tram from Fulwell to Hammersmith – one of the rare migrations of such cars.

W. H. Shaw, General Traffic Superintendent of the Underground Group Tramways, emphasized the high traffic earning potential of the Felthams in his paper 'Traffic Experiences with Modern Tramcars', given at the congress.

The future for tramway modernization still looked bright. Frank Pick had said that people should not blame trams for narrow roads. 'We are convinced that even [sic] trams have a life before them.'

At the Underground Group meeting in February 1932 Lord Ashfield said that the tramway modernization policy was wise. Their tramways were carrying more traffic in 1931 than in 1930. Unfortunately on the Uxbridge Road route factories had laid off men because of the economic state of the country, so that the increase in receipts from the Felthams was less than hoped for. The trolleybuses had increased traffic and reduced costs sufficiently to justify the capital cost.

Ministry of Transport regulations dated May 26, 1932, for the LUT trolleybuses show that they were limited to a top speed in service of 25 m.p.h. with a 15 m.p.h. general restriction through Kingston and Teddington. Under all bridges there was a 10 m.p.h. restriction, with a 5 m.p.h. limit under Kingston Road railway bridge in Malden which had not then been rebuilt.

Top speed round turning circles was limited to 5 m.p.h. and this also applied to the passage through all wire junctions and crossings, round right angle bends of intersecting streets, in Kings and Park roads, Kingston, at Malden Fountain cross roads and at the junction outside Surbiton station. There were some compulsory stops.

Finding the trolleybuses heavy at first on current, Spencer fitted two with supersaturated motors, one by BTH, the other by English Electric. In fourteen months' service they cut consumption at starting and in acceleration by more than 30 per cent and per car mile by more than 15 per cent.

The opening of the Kingston By-Pass in 1925 had stimulated building south of the Tolworth tram terminus. To tap the new estates the LUT applied to the Ministry of Transport in 1932 for a Provisional Order to extend from the 'Red Lion' by a loop via Ewell Road, a by-pass service

LONDON UNITED TRAMWAYS LIMITED.

INTERIOR VIEW OF CAR.

Special Saloon Car
To Seat 20.

Rate.—7/6 Return for each 1d. of the Ordinary Fare.

Example:
TWICKENHAM TO HAMMERSMITH.
Single Fare by Tram 4d.
Return Fare for the complete party by Special Car £1 10 0

This Car can be engaged by parties visiting theatres, concerts, balls and other social festivities.

Top: One of the two special cars for private use, at Chiswick Depot. (*London Transport*). *Centre:* The Pullman-type saloon of one of the private hire cars. (*London Transport*). *Bottom:* No. 342, type S2, rebuilt from private hire saloon for PAYE, one-man operation, at Hanwell Depot in 1924. (*London Transport*)

Left: No. 341, the first PAYE car, in Boston Road.

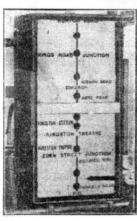

Above: Moving route indicator inside PAYE car on Tolworth–Richmond Park Gates route. (*Courtesy G. L. Gundry*). *Right:* Vestibule of PAYE car No. 342. (*R. B. Parr.*) *Below:* Type S2 No. 344 prepares to leave Hanwell for Brentford. (*G. L. Gundry*)

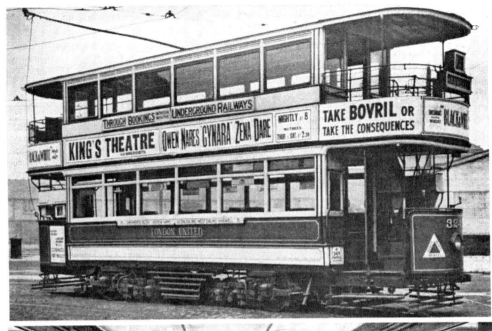

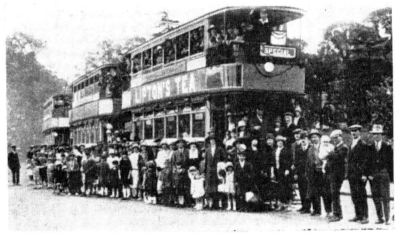

Top: Type T No. 324 after remotoring and fitting of transverse seats in lower saloon in 1925. (*London Transport*). *Centre:* Interior of top deck of type T car No. 307 after reconditioning and fitting of transverse upholstered seats. (*London Transport*). *Bottom:* The Hounslow Club and Institute (LUT) wives' and children's outing by tram to Barnet, for Hadley Wood, via Acton and Cricklewood, in 1925. (*Courtesy G. L. Gundry*)

Above: Lone 'Poppy' No. 350 at Gunnersbury. Note the similarity of outline of this LGOC-designed car of 1929 to the LGOC 'NS' bus behind. *Centre:* Type T No. 320 at Kew Bridge in 1933. (*Courtesy R. B. Parr*). *Below:* No. 261, seen at Shepherds Bush in 1928, was originally type W and then type U. It became type WT in 1928 when a type T top cover with deep end panels was fitted.

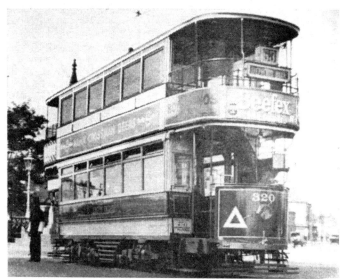

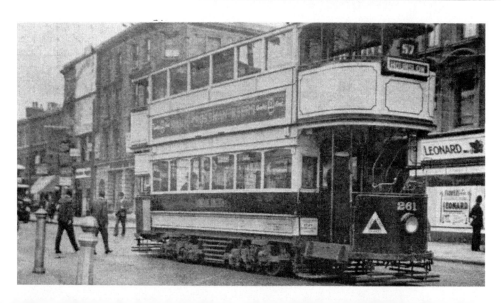

Finale

road and Warren Drive, adding another 1,500 yards to the 17¼ miles already converted.

The cost was put at £11,000, vehicles included. A novelty in construction was the use of concrete standards. The extension was opened by the new London Passenger Transport Board on Wednesday September 20, 1933.

In announcing the terms for the acquisition of the tramways by the proposed London Passenger Transport Board, Lord Ashfield spoke of the 'exceedingly difficult problem' of the MET and LUT which, after full allowance had been made for depreciation and obsolescence, had lost on their operations for years.

The Board entered into its heritage on January 1, 1933. On July 1st it took over the LCC tramways and all the municipal and company tramways in London and the suburbs.

LUT passenger vehicles taken over totalled 150 tramcars (9 original open top, 45 covered, 9 extensive rebuilds, 40 T class, the experimental car 'Poppy' of 1928 and the 46 Felthams), 60 rear-entrance trolleybuses and a 74-seat centre-entrance trolleybus put into traffic on March 2, 1933.

The 74-seater seemed to be the much-desired trolleybus of tramcar capacity. It was a joint product of the Underground Group, AEC and English Electric. Novel features included its near-centre entrance with power-worked doors under control of the driver. Motor and control equipment were underfloor. It entered service in May 1933 on the Wimbledon-Hampton Court route.

This vehicle and the rapid development of trolleybuses in general soon made the original LUT trolleybuses look outdated, though they performed excellent service right through the second world war. Most were withdrawn in 1948, though a few lasted longer as staff training vehicles. No. 1 was preserved and is housed in the British Transport Museum at Clapham, sole survivor of the LUT fleet.

From the tone of Lord Ashfield's address to the Royal Society on 'London Passenger Transport and Street Congestion' in April 1933, it was clear that the new Board had hardened its heart against tramways. They were now seen as an undoubted cause of congestion, forcing other traffic to stop while the cars loaded and unloaded. Their termini at important junctions took up valuable road space just where it was most needed.

It would have been fairer if Lord Ashfield had explained the reasons why the London tramways had suffered under such disabilities. The many stub termini he complained of, as well as the varying street widths, were admittedly a great handicap and were a factor which contributed to their ultimate disappearance in London. But they were not intrinsic defects. Indeed the LUT had pointed the way with its own Hammersmith loop.

On October 12, 1933, Spencer resigned the General Managership of what had become the Board's Northern and Western tramways to become Resident Director of the North Metropolitan Electric Power Supply Company. T. E. Thomas, General Manager until July 1st of the LCC tramways and after that of the Board's Central, Southern and Eastern tramways, became General Manager for all trams and trolleybuses.

Thomas had hardly taken office when the LTPB promoted a Bill to convert no fewer than ninety more miles of tramways in west, north-west and south-east London.

The proposed conversion covered all remaining LUT tram routes. It seems probable that as unified control offered a new field of operation for the Feltham cars south of the river on the busy ex-LCC system, the new Board had no qualms about making the sweep complete.

The Hammersmith/Shepherds Bush–Hounslow/Hampton Court routes were converted on October 27, 1935, the Hammersmith–Acton section on April 5, 1936. The Uxbridge route followed on November 15, 1936 and the Brentford–Hanwell on December 13, 1936, after which the Feltham cars were transferred to the former LCC depots at Telford Avenue and Brixton Hill. There they were joined in 1938 by the former MET Felthams, displaced by trolleybus conversion in their turn. All the fleet thereafter worked services 8/20, 10, 16/18 and 22/24, until being sold in 1949 to Leeds City Transport, where they survived until 1959.

Epilogue

What now remains of the London United? Of the fleet, one survivor, 'Diddler' No. 1, preserved in pristine condition at the Museum of British Transport. Alas, LUT Feltham No. 369, which became Leeds No. 517 and was finally acquired by the Middleton Railway Trust, was burned for scrap in April 1968 after languishing in a field for nine years. As some consolation MET Feltham No. 355 keeps No. 1 company at Clapham.

The depots and substations survive as substantial monuments, some shorn of former glory. Hanwell depot alone is still in full transport use, as an operational London Transport bus garage. Part of Fulwell, too, is so used, but the rest is leased to a local manufacturer for light industrial and storage purposes.

The sheds at Chiswick are leased to British European Airways for the operation and maintenance of coaches serving London Airport. The power house, latterly Goldhawk substation, is a store for London Transport engineering equipment. It is satisfactory to know that the splendid iron stairs and gallery are intact and that the decorative weight-driven clock still ticks.

Acton depot is a cable store of London Transport's H.T. Mains Engineer and its forecourt serves as a bus stand and terminal. Isleworth depot is now used by the Post Office, which took it over in 1968 after the reconstruction of Hounslow Garage had made it redundant as a bus garage. The 'white elephant' Hillingdon, or Hayes, depot, which has known a number of changes, is now owned by a firm.

The substation at Kingston was first reconstructed as a maintenance depot for buses but since 1960 has been owned by an engineering firm. That at Wimbledon is a store for London Transport engineering equipment.

At Fulwell and some other places some running rails and pointwork remain visible. Much street track has been lifted but some is still in place, tarred over, to be revealed once more, from time to time, when road works are undertaken.

A surprising number of poles remain, doing duty as lighting standards. One known to the author, in Francis Grove, Wimbledon, still carries its ornamental base.

Happily, many former LUT men are still with us to recall their old company with affection. The continued cameraderie owes much to the flourishing LUT Athletic Club at Fulwell, one of the sports groups of London Transport and direct descendant of the association of Robinson's day. It keeps alive not only the acknowledged sporting prowess of LUT men but also the company's name and spirit of service. Long may it do so!

Though few of them probably realize it, motorists and other road users in the former LUT-served districts have cause to bless the memory of the LUT and its shareholders for fortuitously bequeathing them numerous road and bridge widenings which might otherwise have been long deferred, or indeed never have been carried out.

Many a terrace of houses or row of shops can be dated from the coming of the LUT, though in some places some pretty scrappy ribbon development resulted. On the other hand the LUT certainly contributed to the rise of shopping centres, particularly Kingston.

Garrick Villa has been restored and divided into expensive flats. Equally expensive modern terraced houses occupy part of the grounds. The course of the tram siding was visible until recent years but road works have since obliterated it. The estate is now called Garrick's Village.

When one considers the weight of opposition, the wonder is that the London United ever got going at all. Only a Robinson could have done it. Indirectly the LUT can be said to have fostered the MET, as Robinson's ambitions certainly goaded Middlesex County Council to emulation. The ifs and buts of transport history are a fascinating, if idle, speculation. In view of the swift advent of reliable motorbuses it is perhaps as well that the LUT was unable to inflate itself to the full extent of Robinson's dreams. As it was, there were too many thin patches. With our hindsight we can see that in the early years too little was placed to renewals, to the later detriment of cars and track. By the time that financial reorganization began to yield results, much ground had been lost, though Spencer did wonders.

The wisdom of partial trolleybus conversion, as carried out in LUT days, cannot seriously be doubted. It is debatable whether reserved-tracking of part of the Uxbridge route would have done more than defer the evil day. The abandonment of miles of reserved track in provincial cities is proof enough. Even if it had been possible to upgrade the Uxbridge route, how long would it have remained economic in an otherwise largely trolleybus system and separated from those few sections of the MET which might have been candidates for similar treatment?

Appendix I

Routes and Service Numbers

BY A. W. MCCALL

The four routes operated in 1895 were:

(a) Shepherds Bush–Askew Arms (dark brown cars) every nine minutes. Fare 1d. Later in the year when track relaying had been completed through Acton, and the extension there opened, the route was extended to Acton Hill ('White Hart')
(b) Shepherds Bush–Young's Corner (yellow cars) every six minutes. Fare 1d. On Sundays and public holidays the route was extended to Kew Bridge
(c) Hammersmith Broadway–Kew Bridge (chocolate cars) every five minutes. Fare 2d
(d) Kew Green–Richmond (Church Road) (yellow cars) every 12 minutes. Fare 1d.

The electric era began on Thursday April 4, 1901, when the first public service electric trams in London began running from both Shepherds Bush and Hammersmith to Kew Bridge, and from Shepherds Bush to Acton, leaving only the Richmond section still operated by horse cars.

When the extension from Kew Bridge to Hounslow ('Bell') opened on Saturday July 6, 1901, services were provided from both Shepherds Bush and Hammersmith. On Wednesday July 10, 1901, the extension of the Acton route to Ealing, Hanwell and Southall Town Hall opened. As at January 1, 1902, the following services were operated:

Electric: Shepherds Bush–Hanwell and Southall
Shepherds Bush–Kew Bridge
Shepherds Bush–Hounslow (Bell)
Hammersmith–Hounslow (Bell)
Hammersmith–Kew Bridge
Horse: Kew Green–Richmond (Church Road)

Appendix

On Wednesday August 13, 1902, the two Hounslow routes were prolonged over the new extension to the 'Hussar', on the corner of Barrack Road, on the eastern edge of Hounslow Heath. Although the extension added a mile to the routes the fares remained as when the cars terminated at the 'Bell'.

On the same day the first section of the Thames Valley lines was opened, from Busch Corner on the Hounslow route via Twickenham Road, Isleworth, and London Road to Cole's Bridge, Twickenham (the present Hill View Road fare stage). To work this section, two new routes were provided, from Shepherds Bush and Hammersmith.

On Saturday September 13th the Cole's Bridge section was extended to the junction of Cross Deep and King Street in Twickenham and the services were likewise extended. On the same day the branch was opened from the junction of London Road and York Street, Twickenham, to Cambridge Road, on the west side of Richmond Bridge and a shuttle service began running between Cross Deep and Richmond Bridge.

On November 7th the extension from Cross Deep to the 'Nelson' (Stanley Road Junction) was opened and the Richmond Bridge and Shepherds Bush services were prolonged to the new terminus, the service from Hammersmith still terminating at Cross Deep.

As at January 1, 1903, the route list was as follows:

Shepherds Bush–Hanwell and Southall
Shepherds Bush–Kew Bridge
Shepherds Bush–Hounslow Heath
Shepherds Bush–Stanley Road Junction
Hammersmith–Kew Bridge
Hammersmith–Hounslow Heath
Hammersmith–Cross Deep
Richmond Bridge–Stanley Road Junction
Kew Green–Richmond (Church Road) (horse)

On April 2, 1903, the section between Stanley Road Junction and Hampton Court via Teddington and Hampton Wick was opened and served by an extension of the Richmond Bridge–Stanley Road route. On April 4th the western arm of the loop, via Hampton Hill to Hampton Court, followed suit and the Shepherds Bush–Stanley Road and Hammersmith–Cross Deep services were both extended to Hampton Court.

Appendix

The year 1904 began with the following routes running:

Shepherds Bush–Hanwell and Southall
Shepherds Bush–Kew Bridge
Shepherds Bush–Hounslow Heath
Shepherds Bush–Hampton Court via Hampton Wick
Hammersmith–Kew Bridge
Hammersmith–Hounslow Heath
Hammersmith–Hampton Court via Hampton Hill
Richmond Bridge–Hampton Court via Hampton Hill
Kew Green–Richmond (Church Road) (horse)

The Askew Road and Southall–Uxbridge sections were both inaugurated on Wednesday June 1, 1904, with a new route, operated daily, between Hammersmith and Uxbridge via Acton. Another new service began on June 5th, running on Sundays only between Shepherds Bush and Uxbridge in addition to the daily service between Shepherds Bush and Southall.

The following route list applies for both January 1905 and 1906:

Shepherds Bush–Hanwell and Southall (daily)
Shepherds Bush–Uxbridge (Sundays only)
Shepherds Bush–
- Kew Bridge
- Hounslow Heath
- Hampton Court

Hammersmith –
- Southall and Uxbridge (via Acton)
- Kew Bridge
- Hampton Court
- Hounslow Heath

Richmond Bridge–Hampton Court (via Hampton Wick)
Kew Green–Richmond (Church Road) (horse)

Three of the Surrey routes were opened for traffic on Thursday March 1, 1906:

Hampton Wick junction–Kingston Bridge–Clarence Street–London Road–Kingston Hill ('George and Dragon')
Kingston (Eden Street junction)–St James's Road–Penrhyn Road–Claremont Road–Surbiton station–Victoria Road–Brighton Road–Portsmouth Road–Windows Bridge (Winters Bridge, Dittons)
Surbiton station–St Marks Hill–Ewell Road–Tolworth ('Red Lion')

Appendix

An LUT Notice dated February 24, 1906, published in the *Richmond and Twickenham Times*, gives the following as the routes to be provided for the Kingston area:

Richmond Bridge–Windows Bridge: from Richmond Bridge at 7.45 a.m. then every ten minutes until 9.55 p.m., then at 10.15, 10.40 p.m., then every 20 minutes until 11.20 p.m. From Windows Bridge at 7.15 a.m. then at every ten minutes until 10.50 p.m., then cars every 20 minutes to Fulwell Depot until 12.20 a.m.
Kingston Hill–Windows Bridge: from Kingston Hill at 7.30 a.m., then every 10 minutes until 9 p.m., then every 20 minutes until 11.20 p.m., then to Fulwell until midnight
Kingston Hill–Hampton Court: from Kingston Hill at 7.35 a.m., then every ten minutes until 9.55 p.m., then at 10.10 p.m. and every 20 minutes to 11.30 p.m. Then cars to Fulwell until midnight
Surbiton Station–Tolworth: from Surbiton at 7.20 a.m., then every five minutes until 9.30 p.m., then every 10 minutes until 11.50 p.m.

On May 26, 1906, more lines in the Kingston area were ready for traffic, as follows:

Eden Street Junction–Ham Boundary via Richmond Road
Richmond Road (Kings Road junction)–Richmond Park (Kingston Gate), via Kings Road
Norbiton Church–Malden Fountain via Cambridge and Kingston roads

The opening of these lines caused a rearrangement of services and the *Richmond and Twickenham Times* for May 26, 1906, carried an LUT Official Notice quoting the following routes to start operation on May 26th:

Malden–Hampton Court: 7.25 a.m. then every ten minutes to 9.45 p.m., then every 20 minutes until 11.5 p.m. on weekdays. The Sunday service started at 9.25 a.m., running every 10 minutes until 9.45 p.m., then every 20 minutes until 11.5 p.m.
Ham Boundary–Tolworth: 7.30 a.m. then every 10 minutes to 8.30 p.m., then at 8.50, 9.10 p.m., then every 20 minutes to Fulwell only. On Sundays at 9.30 a.m. then as weekdays
Tolworth–Richmond Park Gates: From Tolworth at 7.15 a.m. then every 10 minutes until 9.5 p.m., then every 20 minutes until 11.25 p.m.

Appendix

In the reverse direction, after the last car to Tolworth at 10.55 p.m., cars ran to Fulwell until 11.55 p.m. On Sundays the first car left Tolworth at 9.35 a.m., the service being every 10 minutes until 9.5 p.m., then every 20 minutes until 11.5 p.m.

At the same time the Kingston Hill–Windows Bridge service was withdrawn south of Surbiton Station except during peak hours.

The reader perusing this list of services may gain the impression that the London United had over-provided for the Kingston area but it will be remembered that the district covered by the new lines was largely built up even by then and, as far as can be ascertained, the only daily bus service still operating was between Kingston and Kew Bridge via Richmond.

The Brentford (Half Acre)–Hanwell route along Boston Road also opened on May 26, 1906. At first it carried both a shuttle service between the points named, the Hanwell terminus being in Boston Road by the Broadway junction, and also a through service from Hammersmith to Uxbridge via Brentford and Hanwell which traversed Lower Boston Road to reach Uxbridge Road. The second service was later cut back to Southall and disappeared about 1910, after which the Lower Boston Road section was no longer in regular use.

The year 1907 began with the following routes operating:

Shepherds Bush–Hanwell and Southall (daily)
Shepherds Bush–Uxbridge (Sundays only)
Shepherds Bush–Kew Bridge
Shepherds Bush–Hounslow Heath
Shepherds Bush–Hampton Court
Hammersmith–Acton–Hanwell–Southall–Uxbridge (daily)
Hammersmith–Brentford–Southall–Uxbridge
Hammersmith–Kew Bridge
Hammersmith–Hounslow Heath
Hammersmith–Hampton Court
Richmond Bridge–Windows Bridge
Richmond Bridge–Hampton Court via Hampton Wick
Malden–Hampton Court
Kingston Hill–Surbiton Station (workmen's cars to Windows Bridge)
Ham Boundary–Tolworth
Richmond Park Gates–Tolworth
Brentford–Hanwell Broadway

Appendix

Kew Green–Richmond (Church Road) (horse cars)

The Malden–Raynes Park Station extension was opened on Saturday April 27th and served by an extension of the Hampton Court–Malden route.

On Thursday May 2nd followed the further extension along Worple Road to Wimbledon Hill Road. The system was rounded off on June 27th by the opening of the lines from Wimbledon Hill Road to the London County Council boundary at Longley Road, Tooting; the loop in Wimbledon via St Georges Road and Francis Grove; the short spur in Wimbledon Hill Road to the foot of Wimbledon Hill; and the Haydons Road branch from Merton to Summerstown ('Plough').

The Hampton Court route was extended to Longley Road terminus, shown on LUT cars as 'Tooting'. The LCC terminus a few yards north appeared as 'Merton', which was wider of the mark.

Initially cars ran between Wimbledon Hill and Summerstown but the service was subsequently cut back to a Merton High Street–Summerstown shuttle, with transfer fares to the Hampton Court–Tooting route.

Baedeker's Guide to London for 1908 quotes two routes as operating from Tooting, each providing a ten-minute service, one to Hampton Court, the other to Richmond Bridge via Teddington.

About this time there was another reorganization of the Kingston routes to achieve better traffic results. It resulted in the following changes:

The Ham–Tolworth service was curtailed to run as a short shuttle between Ham Boundary and Kings Road Junction, with through transfer fares to the Richmond Park Gates–Tolworth route. The Kingston Hill–Surbiton Station route, which had some workmen's cars extended to the Dittons (Windows Bridge) was extended to the Dittons as a daily service. The Richmond Bridge–Tooting service was cut back to Eden Street Junction and the Hampton Court–Tooting service strengthened. To compensate for the loss of the Richmond Bridge service in Wimbledon, the Summerstown–Merton service was extended to Raynes Park, although, it is believed, at rush hours only and at weekends.

The Kew Road horse car service was withdrawn on April 20, 1912.

The London & Suburban Traction Co. Ltd, which assumed control of the LUT and MET on November 1912, decided to number the routes of both companies in a common system, as follows:

Appendix

The LUT was to use odd numbers, except where these were required by the MET for through running services with the Northern Division of the LCC, which also used odd numbers. The MET was to use even numbers, except as quoted above.

The provisos explain the reason for the gap between 7 and 55 on the LUT and the non-use of 59 and 79, but do not explain why the LUT series started at 7, nor why 9 was used for the Hammersmith services to the Uxbridge Road.

The list of service numbers used by the LUT appears below, with the fares and running times at the time when numbers were introduced:

Route No.	Route	Days Operated	Fare	Journey Time (mins.)
7	Uxbridge–Shepherds Bush	Daily	5d	76
9	Uxbridge–Hammersmith (via Acton)	Suns	5d	79
9A	Southall–Hammersmith (via Acton)	Suns	3d	47
9B	Hanwell–Hammersmith (via Acton)	Daily	2d	40
55	Hanwell–Brentford	Daily	1d	13
57	Hounslow Heath–Shepherds Bush	Daily	4d	52
61	Hounslow Heath–Hammersmith	Daily	4d	50
63	Kew Bridge–Shepherds Bush	Daily	2d	21
65	Hampton Court–Shepherds Bush	Daily	6d	70
67	Hampton Court–Hammersmith	Daily	6d	64
69	Richmond Bridge–Kingston	Daily	2½d	38
71	Hampton Court–Tooting	Daily	4d	58
73	Kingston Hill–Dittons	Daily	2d	26
75	Ham Boundary–Kings Road	Daily	1d	3
77	Tolworth–Richmond Park Gates	Daily	2d	27
81	Summerstown–Haydons Road Junc.	Daily	1d	8
83	Hanwell–Shepherds Bush	Daily	2d	33
85A	Kew Bridge–Hammersmith	Daily	2d	19
87	Southall–Shepherds Bush	Daily	3d	44

The first alteration to this list took place in September 1914. The MET and LCC introduced through running between Barnet and Moorgate. One of the routes involved was LCC No. 9, which was extended from Highgate (Archway Tavern) to Barnet. As the MET also participated in the operation of route 9, the LUT routes 9, 9A and 9B were renumbered 89, 89A and 89B to avoid confusion.

Appendix

At first the 1914–18 war caused no alteration to the route network but by the winter of 1916–17 Fulwell Depot was finding difficulty in maintaining all services. In March 1917 Kingston Corporation was notified that it was proposed to discontinue temporarily the working of the Ham Boundary and Kingston Hill sections with effect from March 31st. In the absence of evidence to the contrary it is presumed that route 73 from the Dittons was then cut back to Eden Street Junction.

At the Board meeting on May 1st it was decided to reinstate 73 to Kingston Hill, but the date when service was resumed is not known. The Ham Boundary service was restored on October 29th to provide additional transport for men employed at the Hawker aircraft works being erected near the terminus.

It would appear that the issue of the excellent map and guide by the London & Suburban was suspended during the war years but a route folder dated August 1919 has been seen. It does not quote fares, running times and service intervals, but shows the following routes operated:

- 7T   Shepherds Bush–Uxbridge
- 55T   Hanwell–Brentford
- 57T   Shepherds Bush–Hounslow Heath
- 61T   Hammersmith–Hounslow Heath
- 63T   Kew Bridge–Shepherds Bush
- 65T   Hampton Court–Shepherds Bush
- 67T   Hampton Court–Hammersmith
- 69T   Richmond Bridge–Kingston (Eden Street)
- 71T   Hampton Court–Tooting
- 73T   Dittons–Kingston Hill
- 75T   Ham Boundary–Kings Road (Kingston)
- 77T   Tolworth–Richmond Park Gates
- 81T   Summerstown–Haydons Road Junction
- 83T   Hanwell–Shepherds Bush
- 85A   Kew Bridge–Hammersmith
- 87T   Southall–Shepherds Bush
- 89T   Hammersmith–Birch Grove (Acton) – extended on Sundays to Hanwell

The only wartime casualty was the Sunday operation of 89 and 89A to Southall and Uxbridge.

No map and guide has been seen for 1920. It may be that, like the LGOC, the only L & ST publicity on routes for that year was the maps

Appendix

published at intervals in the national press. One of these has been inspected. It shows the extent of the network and the fare stages but gives no route details.

Pocket maps and guides were resumed in 1921 with an issue entitled 'Summer Services–May to October'. An examination of this reveals the following alterations:

- 55   extended to Ealing – daily
- 57   Saturdays and Sundays only
- 65   Saturdays and Sundays only
- 75   withdrawn
- 85A renumbered 85
- 89   withdrawn between Hanwell and Acton

It is believed that about this time some cars ran through between Richmond Bridge and Summerstown and that there were occasional services to Ham for aircraft workers throughout the 1920s.

The following table lists services operating in the summer of 1921:

		Journey time (mins)	Interval wkdy	Suns	Fare
7	Uxbridge–Shepherds Bush	81	12	8	1s
87 (7A after 1924)	Southall–Shepherds Bush	45	6	4	7d
83 (7C after 1924)	Hanwell–Shepherds Bush	33	3	4	6d
55	Hanwell–Brentford	13	10	10	2d
57	Hounslow–Shepherds Bush	51	12 (Sats)	10	8d
61*	Hounslow–Hammersmith	49	6	10	8d
63	Kew Bridge–Shepherds Bush	21	4	4	4d
65*	Hampton Court–Shepherds Bush	70	12 (Sats)	10	11d
67	Hampton Court–Hammersmith	68	6	10	10d
69	Kingston–Richmond Bridge	38	7½	7	6d
71	Hampton Court–Tooting	60	6	5	9d
73	Dittons–Kingston Hill	28	15	10	3d
77	Tolworth–Richmond Park Gates	28	7½	10	3d
81	Summerstown–Haydons Road Junction	8	20	20	2d
85	Kew Bridge–Hammersmith	19	3	3½	3d
89	Acton–Hammersmith	22	4	5	3d

* discontinued late 1922

Appendix

The next issue of the map was made to cover the summer programme for 1922 and the following alterations are noted:

- 55 withdrawn between Hanwell and Ealing
- 57 reinstated on Mondays – Fridays and runs daily
- 61 withdrawn
- 63 extended to Brentford End on weekdays
- 65 withdrawn except for Bank Holiday weekends
- 71 withdrawn between Wimbledon and Tooting
- 85 withdrawn

It is difficult some forty-five years later to see the reason for some alterations, particularly the withdrawal of 61 completely and 65 partially, especially as there was no bus competition west of Kew Bridge on those two routes. The extension of 55 to Ealing was a praiseworthy attempt to provide a through connection between the developing Boston Manor area and the shops of West Ealing and the amusement centre of Ealing itself.

Route 71 was cut back from Tooting to the Wimbledon loop after the last journeys on May 1, 1922, its place being taken by the extension of LCC routes 2 and 4 to Wimbledon Hill, at last placing Wimbledon on a through tram route to central London. The car operating the now detached Summerstown route 81 continued to use the Merton–Wimbledon section on its depot run to and from Fulwell.

Route 85 was also operated finally on the same date. On May 2nd LCC route 26 from Hop Exchange was extended from Hammersmith to Kew Bridge and the LCC took over Chiswick Depot, whose LUT cars were transferred to Hounslow.

The winter 1922 edition of the map shows further alterations, this time to routes from Hounslow and Fulwell depots.

The long-unprofitable section of route 57 between Hounslow ('Bell') and Hounslow Heath ('Hussar') was withdrawn on Wednesday July 5, 1922. No compensatory service, other than the infrequent journeys on LGOC routes 117 and 117A, was provided for a few days, but on July 12 LGOC service 37A was extended from Hounslow Garage to the 'Hussar' and special early morning journeys were made between Hounslow Heath and Isleworth Station on route 117 for workers at Pears soap factory.

Route 69 was withdrawn and as an experiment 67 was extended from Hampton Court to Eden Street Junction and 71 was diverted via Hampton Wick and Teddington to run to Richmond Bridge.

This experiment was evidently not a success, as the summer 1923 map

Appendix

shows 67 and 71 based on Hampton Court again, and 69 reinstated from Richmond Bridge to Eden Street. In addition 63 was working Shepherds Bush–Brentford (Ealing Road) on weekdays, terminating at Kew Bridge on Sundays.

In the winter programme of 1923, route 63 was withdrawn between Kew Bridge and Brentford and 85 reinstated, this time from Hammersmith to Brentford (Ealing Road) on weekdays only. The reason for this change is obscure, especially as 85 had been withdrawn eighteen months before when LCC route 26 was extended to Kew Bridge.

There must have been a build-up of traffic between both Shepherds Bush and Hammersmith and Brentford, as the summer 1924 map reveals not only 85 still running but 63 once more extended to Brentford (Ealing Road). As the Ealing Road junction was at the narrowest part of the High Street the congestion caused by tramcars reversing must have been considerable – a four-minute service operated on 63 and a six-minute on 85.

The other alterations for the summer of 1924 concerned the routes from Hanwell depot. Up to this time bus operation on the Uxbridge Road had been confined to the section between Shepherds Bush and Ealing Broadway served by LGOC route 17. By 1924 the LGOC had been joined by several independents, including the large fleet of the Cambrian Coaching and Goods Transport Company.

As a member of the Underground Group the LGOC paid due regard to the traffic returns of its sister tram company, and had not worked west of Ealing for many years. The independents had no such scruples and were operating as far as Southall Town Hall, one or two more adventurous spirits running to Hayes Station.

The LUT decided to meet the competition with a slight reorganization of routes. No. 83 to Hanwell was withdrawn, 87 to Southall was renumbered 7A and its service interval cut from six to three minutes. A new route, 7B, was introduced on Saturdays only between Shepherds Bush and Hayes (Yeading Lane), with a six-minute service.

After all the alterations since 1921 it may be appropriate to list routes operated in the summer of 1924:

7	Shepherds Bush–Uxbridge	daily	every	10	mins	W
			every	6	mins	S
7A	Shepherds Bush–Southall	daily	every	2½	mins	W
			every	3	mins	S

Appendix

7B	Shepherds Bush–Hayes	Sats	every	6	mins	
55	Hanwell–Brentford	daily	every	10	mins	
57	Shepherds Bush–Hounslow	daily	every	8	mins	W
			every	6	mins	S
63	Shepherds Bush–Brentford	wkdys	every	4	mins	
63	Shepherds Bush–Kew Bridge	Suns	every	3	mins	
67	Hammersmith–Hampton Court	daily	every	6	mins	W
			every	5	mins	S
69	Richmond Bridge–Kingston	daily	every	7	mins	
71*	Wimbledon–Hampton Court	daily	every	7	mins	W
			every	5	mins	S
73†	Kingston Hill–Dittons	daily	every	10	mins	
77‡	Richmond Park Gates–Tolworth	daily	every	7½	mins	W
			every	10	mins	S
81	Summerstown–Merton	daily	every	20	mins	
85	Hammersmith–Brentford	daily	every	6	mins	W
			every	5	mins	S
89	Hammersmith–Acton	daily	every	3	mins	

* extra cars in peak hours Wimbledon–West Barnes Lane

† additional service on weekdays between Eden Street and the Dittons, making a joint 7½-minute service

‡ extra cars on weekdays Eden Street–Tolworth and daily Eden Street–Surbiton station, making a 3¾-minute joint service between those two points.

In addition, between Hammersmith and Kew Bridge there were also the LCC cars on route 26 from Hop Exchange and although LUT route 65 was not shown on the map it was operated on Bank Holiday Sundays and Mondays between Shepherds Bush and Hampton Court.

On October 1, 1924, the section of 69 between Richmond Bridge and York Street junction, Twickenham, was withdrawn and covered by extension of LGOC route 27. Through tickets were issued between the trams and buses – they included through fares to and from the centre of Richmond.

In the summer of 1925 route 7B was withdrawn as a regular service and thereafter operated at odd times, mainly during weekdays peak hours. Route 83 (Shepherds Bush–Hanwell), which had been withdrawn the preceding summer was reinstated but renumbered 7C.

Route 85 had disappeared by the winter of 1925 but headways on 67

Left: Type T No. 335 in ultimate condition, with windscreens, in early London Transport days (LPTB No. 2351). Note trolleybus overhead for conversion of Uxbridge route in 1936. (*Courtesy N. D. W. Elston*). *Centre:* No. 211, originally type W, became type WT when rebuilt in 1928 with T type top cover and deep end panels. Here seen in Chiswick High Road in early London Transport days as LPTB No. 2408. (*Courtesy S. G. Jackman.*) *Bottom:* No. 396, the last of the Felthams, emerges from Fulwell Depot on 14 February 1931. (*C. F. Klapper Collection*)

Top: Feltham type No. 353 in Ealing Broadway in January 1931. (*London Transport.*) *Centre:* Saloon of a Feltham car, looking towards front exit. (*London Transport*). *Right:* Feltham No. 356 at Uxbridge terminus in April 1933. (*Courtesy M. J. O'Connor*)

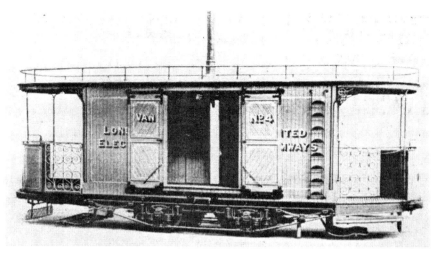

Above: Four-wheel ticket van No. 4 (later 004). (*Courtesy R. B. Parr*)

Above and below left: Water car in original and final (London Transport) condition. (*Courtesy R. B. Parr.*) *Below right:* Bogie stores van No. 005 in early London Transport days.

Top: At Twickenham in 1931. Type W No. 254 passes new trolleybuses waiting to inaugurate service to Teddington. (*Courtesy R. B. Parr*). *Centre:* Trolleybus at Twickenham terminus in May 1931. (*London Transport.*) *Bottom:* Trolleybus No. 33 on Twickenham–Tolworth service in Kingston Road, Teddington. (*London Transport*)

Appendix

were strengthened between Hammersmith and Twickenham, serving the dual purpose of replacing 85 and meeting the new demand caused by new housing along Twickenham Road, Isleworth.

During the latter part of 1926, route 63 operated west of Kew Bridge to Brentford during weekday rush hours only and extra cars on 73 between Eden Street and the Dittons were running during weekday rush hours only instead of all day. At the same time, the Eden Street-Tolworth short workings on 77 were withdrawn and the Eden Street-Surbiton cars relegated to a Saturday afternoon and evening service only, although in summer 1926 they ran Sunday evenings also.

In 1927 it was decided that as alterations to services had become few the map would be issued only once a year. Comparing the 1927 map with that for the summer before, the only alteration that shows up is the inclusion of the summer weekend extension of LCC 2 and 4 to Hampton Court which had begun on Saturday May 22, 1926, too late for inclusion in the summer 1926 edition of the map.

The map for 1928 shows further alterations to the pattern of service on routes 73 and 77. The extras from Eden Street to the Dittons were withdrawn, but the service over the whole of 73 was stepped up from a 20-minute interval to a 10-minute interval between cars. On 77 the rush-hour short workings between Eden Street and Tolworth returned during weekday evening rush hours only. The other alteration was to route 89, on which every other car was extended from Acton to Hanwell on Saturday afternoons and evenings. In addition, the headways on most services were increased on Sundays because of a falling-off in traffic.

An interesting weekday extension to 89 began on Wednesday November 28, 1928, when this route was prolonged from Hammersmith to Putney over LCC tracks. The route was jointly operated by LUT and LCC cars, both of whose destination indicators described the Putney terminus incorrectly. The LUT just called it 'Putney'. The LCC showed 'Putney High Street'. In fact the cars continued along Putney Bridge Road to the crossover at Fawe Park Road, the farthest penetration into central London which LUT cars ever achieved. On Saturdays LUT cars worked through only from Hammersmith to Hanwell, not from Putney.

LUT cars were withdrawn for a period in 1930 but reappeared from April 16, 1931, to balance the takeover of the Summerstown route by the LCC.

An undated map was issued for 1929. At the foot of one of the panels

Appendix

it shows the printer's date as '1.1.29'. It carried only three route alterations: 63 extended from Kew Bridge to Isleworth Fire Station during weekday peak hours; short workings on 71 between Wimbledon and West Barnes Lane withdrawn; and the Saturday afternoon and evening operation on route 89 extended from Hanwell to Southall.

Henceforth the map and guide issued by the three associated tramways adopted a style similar in layout to the contemporary LGOC map as far as route information was concerned. The style began with the issue for 1930 which gives the LUT routes as:

7 Shepherds Bush–Uxbridge, daily. Service Shepherds Bush–Hanwell 1–2 min.; Hanwell–Southall 3–4 min.; Southall–Uxbridge 10 min. weekdays, 8 min. Sundays

55 Brentford–Hanwell, daily. Extended to Ealing on weekdays. Service Brentford–Hanwell 8 min. weekdays, 10 min. Sundays; Hanwell–Ealing weekdays only 8 min.

57 Shepherds Bush–Hounslow. Every 8 min. daily

63 Shepherds Bush–Kew Bridge. Every 5 min. daily. Extended to Isleworth Fire Station during weekday rush hours only every 8 min.

67 Hammersmith–Hampton Court, daily. Service weekdays 6 min., Sundays 8 min.

69 Twickenham–Kingston, daily. Service weekdays 6 min., Sundays 8 min.

71 Wimbledon–Hampton Court, daily. Service weekdays 6 min., Sundays 8 min. (every 4 min. jointly with LCC routes 2/4 on summer Saturday afternoons and Sundays)

73 Kingston Hill–Dittons, daily. Every 10 min.

77 Richmond Park Gates–Tolworth, daily. Every 10 min. Augmented between Tolworth and Kingston during weekday rush hours and between Surbiton Station and Kingston on Saturday afternoons and evenings

81 Summerstown–Merton, daily. Every 20 min.

89 Acton–Hammersmith, daily. Extended to Putney on weekdays. Service: Acton–Hammersmith, weekdays every 3 min., Sundays every $4\frac{1}{2}$ min.; Hammersmith–Putney every 8 min. weekdays only

It will be seen that 7A and 7C had disappeared. The route stencils of the Feltham cars, introduced about this time, were rather inaccessible,

Appendix

making it difficult to switch indications from 7 to 7A or 7C in successive journeys as in former practice. All workings on the route were therefore simply numbered 7.

The opening of factories on the Great West Road was attracting workers from West Ealing and Hanwell. This traffic had resulted in the weekday extension of 55 shown in this list, although an amendment leaflet to the map dated June 23, 1930 by the printer states that the extension to Ealing operated during weekday peak hours only.

The map for 1931 was published in the early part of the year. It shows no alterations to services and is the last to show only tramway operation on the LUT.

On Saturday May 16th tram service 69 was withdrawn between Twickenham and Teddington and replaced by trolleybus route 1. Trolleybus 1 was extended on Monday June 15th to operate between Twickenham and the new Kingston Hill loop, entirely replacing tram service 69. As no immediate alteration was made to routes 73 and 77 a peculiar arrangement resulted near Richmond Park Gates. After leaving this terminus tram conductors had to dip the trolley pole to clear the trolleybus wiring at Park Road, where no frogs or crossings were installed. After the tram coasted across, the conductor replaced the pole on the trolleybus positive wire beyond.

This procedure continued until Wednesday July 15th when the Eden Street–Tolworth section became ready for trolleybus operation. Route 1 was diverted at Eden Street to operate between Twickenham and Tolworth ('Red Lion'), and a new route 2 was introduced between the Kingston Hill Loop and Tolworth, working both ways round the loop. Tram 77 operated its last journeys on July 14th and 73 was curtailed to operate between Surbiton station and the Dittons, except for journeys to and from Fulwell Depot. Although not advertised in the maps, short workings by the trolleybuses were operated between Surbiton and Tolworth as route 1A.

On Tuesday July 28th Tram 73 made its last journey. Trolleybus 3 was introduced between the Dittons and the Kingston Hill Loop the next day.

The early operation of trolleybus routes 2 and 3 is of interest. Buses on 2 would start from Tolworth and proceed via Richmond and Kings roads, Kingston. Near Park Road the blind would be altered to '3 Dittons' and the bus would then continue via Park Road and London Road.

Appendix

To cover the Kingston loop in the reverse direction the trolleybuses on 3 would start from the Dittons and enter the loop via London Road and Park Road, where the blinds would be altered to read '2 Tolworth' – the journey continuing via Kings and Richmond roads.

It is not clear when this inter-working ceased as the Underground group tramways do not seem to have issued a map for 1933. The LCC Tramways map for May 1933 still shows LUT 2 as operating to Tolworth and 3 to Dittons, but the first LTPB tram map, issued in November 1933 shows them as 2 to Dittons and 3 to Tolworth. The trolleybus extension loop from Tolworth ('Red Lion') to the Kingston By-Pass was opened by the LPTB on Wednesday September 20, 1933, and the two routes may have changed over then.

The Wimbledon route was first served by trolleybuses on Wednesday September 2, 1931, as route 4. Trams on route 71 were not withdrawn immediately as insufficient trolleybuses had been delivered by the opening date to cover the peak hour and Saturday afternoon service. Some trams were operated for at least two to three weeks after the opening.

When first operated to Wimbledon, trolleybuses turned by using Francis Grove and standing in St Georges Road, departing via Wimbledon Hill Road and Worple Road. This arrangement was counterwise to tramway operation.

On December 15, 1932, the trolleybuses were extended to a new turning circle in front of Wimbledon Town Hall, to which point LCC tram routes 2 and 4 were cut back, the cars terminating in the centre of a triangular island.

The first bus of the day left Kingston for Wimbledon at 5.15 a.m. on weekdays and Saturdays. The first from Wimbledon to Hampton Court left at 5.46 a.m. and the first in the return direction left Hampton Court at 6.15 a.m. Last buses were at 11.40 p.m. from Wimbledon and 11.17 p.m. from Hampton Court. On Sundays the first bus left Kingston for Wimbledon at 8.13 a.m. and the last arrival at Hampton Court was at 12.9 midnight.

Unlike the other members of the Underground group tramways the London United enjoyed only limited through running with adjacent systems. The single-line connection between the LUT and MET at Acton had not been sanctioned for regular passenger service, being mainly used for the transfer of cars for overhaul at Hendon Works.

It was unfortunate, particularly in view of their subsequent close

Appendix

association, that there was no direct physical link between the LUT and SMET.

Connecting trackwork at Hammersmith Broadway and Longley Road and a change-pit at the latter point were laid in the early part of 1922 and the LCC ran trial cars through to the LUT on April 24th. After that date, practice runs were made to familiarize motormen with the route and on May 2nd LCC routes 2 and 4 were extended to Wimbledon Hill and route 26 to Kew Bridge. On March 12, 1926, an LCC car made a trial run over LUT tracks to Hampton Court in preparation for the extension of LCC 2 and 4 thither on Saturday afternoons and evenings, Sundays and Bank Holidays from May 22nd. In winter, LCC cars ran to Hampton Court only on Saturday afternoons and evenings.

On June 8, 1924 (Whit Monday), ten extra cars were placed on LCC route 26, operating between Clapham Junction and Kew Bridge. They operated every Bank Holiday thereafter. On Whit Monday (May 24), 1926, this extra service was extended to run to and from Clapham ('Plough') via Cedars Road, and from August Bank Holiday (August 2nd) 1926 further extended to Camberwell Green via Stockwell, Brixton and Loughborough Junction.

When on November 28, 1928, LUT route 89 was projected from Hammersmith over LCC tracks to Putney, LUT cars operated for the first time over foreign metals. Withdrawn from the joint service for a period in 1930 they reappeared from April 16, 1931, to balance the takeover of the Summerstown service by the LCC; they were finally withdrawn in 1933.

LCC route 14 had been a weekday service between Hop Exchange and Wandsworth High Street, extended to Earlsfield Station in peak hours, with certain journeys, mainly in the morning peak, operating to Summerstown. On April 16, 1931, it absorbed the isolated Summerstown service and became a daily service between Wimbledon and Wandsworth High Street via Haydons Road and Garratt Lane, being extended on weekdays via Battersea and the Embankment to Hop Exchange.

From perusal of LCC maps for the period 1931 onwards it would seem that the operation of LCC 2 and 4 to Hampton Court at weekends must have ceased sometime during August 1931, possibly after the Bank Holiday working.

Another working, of which there is no mention in the route maps of either authority, is that by LCC cars on route 89 between Acton and Tooting Junction on Saturday afternoons and evenings for a short

Appendix

period. It was probably during the 1930 period when LUT cars were not participating in the joint service.

A supplementary working of trolleybus route 4 was introduced to give a service between Teddington (Savoy) and Malden, mainly during peak hours and on Saturday afternoons and evenings. This working became LPTB service 5.

The final list of routes as at July 1, 1933, when the system passed to the LPTB, was:

Tramways

- 7   Shepherds Bush – every 2½–3 min. – Southall – every 4–5 min. – Hayes – every 8–10 min. – Uxbridge. Running time 68 min. Fares: 1s single, 1s 6d return
- 55   Brentford–Hanwell. Every 8–10 min. Running time 13 min. Fare 2d. Extended to Ealing and Acton every 8 min. during weekday rush hours only. Fare 5d. Running time 23 min.
- 57   Shepherds Bush–Hounslow. Every 4–10 min. Running time 40 min. Fare 6d
- 63   Shepherds Bush–Kew Bridge. Every 8–10 min. Running time 19 min. Fare 3d. Extended to Isleworth Fire Station every 8 min. during weekday rush hours. Fare 5d
- 67   Hammersmith–Hampton Court. Every 7–10 min. Running time 60 min. Fares: 10d single, 1s return
- 89   Hammersmith–Acton. Every 3 min. weekdays, every 5 min. Sundays. Running time 19 min. Fare 3d. Extended to Brentford via Ealing on Saturday afternoons. Running time 48 min. Fare 8d
- 65   Shepherds Bush–Hampton Court. Every 10 min. on Bank Holidays only

Tram Services Operated by LCC partly over LUT

- 2/4   Wimbledon–Victoria Embankment. Every 4 min. Running time 55 min. Fares: 5d single, 8d return
- 14   Wimbledon–Hop Exchange. Every 12 min. Running time 78 min. Fares: 5d single, 8d return. On Sundays runs Wimbledon Station–Wandsworth High Street only. Running time 22 min. Fares: 4d single, 6d return
- 26   Kew Bridge–Hop Exchange. Every 6 min. Running time 83 min. Fare 7d single, 1s return. On Bank Holidays service 26Ex

Appendix

      also operated Kew Bridge–Camberwell Green. Fare 7d single, 1s return
89   Putney–Acton. Every 10 min. weekdays only. Running time 34 min. Fares: 5d single, 9d return

Trolleybuses

1   Twickenham–Tolworth ('Red Lion'). Every 5 min. weekdays, 8 min. Sundays. Running time 41 min. Fare 6d
2   Tolworth–Kingston Hill Loop. Every 8 min. weekdays, 12 min. Sundays. Running time 25 min. Fare 3d
3   The Dittons–Kingston Hill Loop. Every 8 min. weekdays, 12 min. Sundays. Running time 21 min. Fare 3d
4   Hampton Court–Wimbledon. Every 5 min. weekdays, 7 min. Sundays. Running time 37 min. Fare 6d single, 8d return

Additional Trolleybus Services (not advertised in map)

1A   Tolworth–Surbiton Station
4   Teddington (Savoy)–Malden Fountain

Appendix II

Fares and Tickets

BY A. W. MCCALL

Most early tramways charged a minimum fare of 1d, which covered a distance of 1½ miles. This applied to the three short routes of the four which the LUT took over in 1895: Shepherds Bush–Acton (1½ miles); Shepherds Bush–Youngs Corner (1¼ miles); and Kew–Richmond (1½ miles). On the Hammersmith–Kew Bridge route, a through fare of 2d was charged for the 3-mile journey. For shorter distances this route was divided into three equidistant overlapping fare sections of 1 mile each, for which the fare was 1d. The three sections were as follows:

Hammersmith Broadway–'The Pack Horse and Talbot' (in Chiswick High Road, opposite Turnham Green Terrace)
Youngs Corner–Gunnersbury Station
Turnham Green Church–Kew Bridge

The Shepherds Bush–Youngs Corner route was extended to Kew Bridge on Sundays and Bank Holidays and the same principle applied as for the Hammersmith route, the exception being that the first penny stage from Shepherds Bush, which covered the full weekday journey to Youngs Corner, was extended to 'The Pack Horse and Talbot', about two miles.

This information was compiled from a list published in 1895 by the London County Council Statistical Department.

Mr M. Gibson of Wolverhampton has, however, seen a 2d and 3d ticket for the Hammersmith–Kew Bridge route, issued around the period of the takeover by the LUT and quoting fares as under:

2d Ticket – White
Hammersmith Broadway–'The Pack Horse and Talbot'
Youngs Corner–Kew Bridge
3d Ticket – Magenta
Hammersmith Broadway–Kew Bridge

Appendix

Mr Gibson has seen no ticket of contemporary style to indicate that there were fares at 1d at that time.

The finances of the West Metropolitan were in a parlous state in the last few years of its existence and it may well have been that 2d was the minimum fare charged on the Hammersmith–Kew route.

The fares were soon reduced by the LUT and the following scale applied:

Hammersmith Broadway
1 Youngs Corner
1 1 Pack Horse and Talbot
2 1 1 Turnham Green Church
2 1 1 1 Gunnersbury Station
2 2 2 1 1 Kew Bridge

Before the advent of electric cars, two of the 1d stages were lengthened, that from Kew Bridge being extended to 'The Pack Horse and Talbot', whilst the stage from Hammersmith was extended to Turnham Green Church, giving a final faretable as follows:

Hammersmith Broadway
1 Youngs Corner
1 1 Pack Horse and Talbot
1 1 1 Turnham Green Church
2 1 1 1 Gunnersbury Station
2 2 1 1 1 Kew Bridge

Considering the large sums which had to be paid to local authorities before gaining assents to extensions of the system, the fares charged in the infancy of the electric era were very reasonable.

From a study of early tickets, it would seem that, instead of an orderly system of fare stages set at equal distances on a mileage scale as used at the present day, the fare stages were set at major traffic points on the route.

Taking the Shepherds Bush–Hounslow route as an example, when this route commenced from Shepherds Bush to Hounslow 'Bell' on July 6, 1901, the following faretable applied:

Shepherds Bush
1 Youngs Corner
1 1 Turnham Green Church
(a) (b) (c)

Appendix

```
(a) (b) (c)
 2  1  1   Kew Bridge
 2  2  1  1   Brentford Half Acre
 3  2  2  1  1   Busch Corner, Isleworth
 3  3  2  2  1  1   Pears' Fountain, Isleworth
 4  3  3  2  2  1  1   Hounslow Terminus
```

Taking the penny fares on the above scale, the mileages are as follows:

Shepherds Bush–Turnham Green Church	2.39
Youngs Corner–Kew Bridge	1.75
Turnham Green Church–Half Acre	1.62
Kew Bridge–Busch Corner	1.75
Half Acre–Pears' Fountain	1.65
Busch Corner–Hounslow Terminus	2.01

When twelve months later this route was extended 1.01 miles to Hounslow Heath, no extra fare stages were inserted, nor was any increase made to existing fares. As a result, the penny fare from Busch Corner was extended to Hounslow Heath, 3.02 miles.

The longest 1d stage on the Uxbridge route was the 2.86 miles from Acton (Horn Lane) to Hanwell Broadway, whilst the shortest 1d fare on the whole system seems to have been the 1.5 miles from Surbiton Station to Tolworth ('Red Lion'). This fare was still 1d in 1939, when the penny carried less value than it did in 1906, when the Tolworth route started.

The throughout fares from Shepherds Bush to the outer terminals are quoted below, with mileages:

Shepherds Bush–Hounslow Heath	8.2 miles	4d
Shepherds Bush–Hampton Court	12.6 miles	6d
Shepherds Bush–Uxbridge	12.0 miles	5d

The impression given is of excellent value for each fare, until one realizes how unfairly short-distance riders were treated. Taking the Uxbridge route as an example, there were only three fare stages and fares from Hanwell Broadway to Uxbridge terminus, as follows:

```
        Hanwell Broadway
 1      Southall Town Hall
 2  1   Hayes Post Office
 3  2  1   Uxbridge, Harefield Road Terminus
```

Appendix

This was very satisfactory for a passenger travelling the full 4.43 miles from Hanwell Broadway to Hayes Post Office, but one joining the car at North Road (the last stop before Southall Town Hall) and wishing to travel to Southall Canal Bridge (Delamere Road), a distance of 1 mile, would also have to pay 2d as the 1d stage ended at Southall Town Hall.

In May 1910 the above situation was alleviated, whether as a result of complaints from passengers or because of a fall-off in the number of short-distance riders carried, is not known.

A revised fare structure was introduced whereby each 1d stage on the old faretable was divided into three equal sections, leaving the through fares unchanged, whilst many of the traditional intermediate fares and all the original fare stages were retained and others added. The addition of these extra stages eliminated the grievances of the short-distance passenger.

The main effect was to bring the fares on to a standard basis of three stages – 1d; four stages – 1½d; six stages – 2d; seven stages – 2½d; nine stages – 3d; ten stages – 3½d; twelve stages – 4d; thirteen stages – 4½d; fifteen stages – 5d; sixteen stages – 5½d; eighteen stages – 6d. The only exception to this scale was on the Uxbridge route where the throughout fare from Shepherds Bush (sixteen stages) and Hammersmith (seventeen stages) remained at 5d.

This adjustment apparently exhausted the company's charging powers. When in the latter part of the first war it was necessary to increase fares to cover rising costs, it had to proceed under the emergency procedure of the following acts; (a) Statutory Undertakings (Temporary Increase of Charges) Act, 1918; (b) Tramways (Temporary Increase of Charges) Act, 1920.

Increases under the 1918 Act were put into operation on December 9, 1918, and were on the basis of 1d single fare for 1 mile (the workmen's single fare to be 3 miles for 1d). The routes were to be divided into fare stages half a mile in length, and children's fares were withdrawn.

Whilst the exact intermediate fares cannot be quoted after this lapse of time, a publicity leaflet issued on July 3, 1919, to encourage pleasure traffic on the Underground group quotes the following through fares (1910 fare shown in brackets):

Route

7	Shepherds Bush–Uxbridge	10d	(5d)
57	Shepherds Bush–Hounslow	6d	(4d)

Appendix

61	Hammersmith–Hounslow	6d	(4d)
65	Shepherds Bush–Hampton Court	10d	(6d)
67	Hammersmith–Hampton Court	10d	(6d)
71	Tooting–Hampton Court	6d	(4d)

The increase of fares under the 1920 Act authorized an increase of ordinary fares by 50 per cent and was introduced on December 6, 1920. The minimum fare for adults became 1½d for two stages; 2d for three stages; 3d for five stages, with fares above threepence at two sections for 1d. The minimum child's fare was 1d for two sections, with higher fares at half the ordinary fare; fractions of a penny counted as a penny, for example the child's fare for 3d and 5d adult fares was 2d and 3d respectively. The throughout fare to Hampton Court remained unchanged. The fare to Hounslow and that from Tooting to Hampton Court increased to 8d. The Shepherds Bush–Uxbridge fare was raised from 10d to 1s.

Some time between December 1920 and the next fare alteration on January 1, 1923, 1d fare for one fare stage of half a mile was introduced, with the 1½d fare remaining at two stages.

On January 1, 1923, a new fare scale was introduced which had the effect of reducing quite a few of the intermediate fares, although the throughout fares remained the same. The scale was: two stages, 1d; four stages, 2d; six stages, 3d; eight stages, 4d; ten stages, 5d; twelve stages, 6d; fourteen stages, 7d; sixteen stages, 8d; eighteen stages, 9d; twenty stages, 10d; twenty-two stages, 11d; and twenty-four stages 1s.

All fare stages were numbered, the number being displayed on a metal plate on the trolley standard carrying the stop flag at the fare stage point. The numbers were also shown on the farebill inside the cars, so that passengers could easily check the fare charged.

This scale was to remain in operation for the remainder of the company's existence. In fact, apart from an increase of the minimum fare from 1d to 1½d in July 1940, it existed until the LPTB's major fare increase in February 1947.

Workmen's Fares

In the company's early days, both single and return workmen's fares existed. On the shorter routes with a fare scale up to 4d ordinary adult fare, there was a throughout scale of 1d single, 2d return. On the longer routes to Hampton Court and Uxbridge, the 1d single and 2d return fare applied to two separate sections as under:

Appendix

London terminus–Twickenham (King Street); Twickenham (King Street)–Hampton Court
London terminus–Southall Town Hall; Southall Town Hall–Uxbridge

The Tooting–Hampton Court route only issued workmen's fares between Tooting and the west end of Kingston Bridge at a fare of 1d single, 2d return for the throughout journey.

These workmen's fares remained unchanged when fares were revised in 1910.

Although the 1918 Act permitted workmen's single fares at 1d for 3 miles, and 1d, 2d and 3d workmen's single tickets of the numerical stage type issued when this increase came into force have been seen, it has not been possible to trace precise details of what was actually charged on each route.

The December 1920 fare increase established the scale for workmen's fares which was to apply for the rest of the company's life, and was also used by the LPTB until February 1947, with the exception that the 2d minimum was increased to 2½d return in July 1940. The scale applied was single fare for the return journey with a minimum of 2d.

Children's Fares

Because of the very long distances allowed for adult fares, there were no children's fares at half-rate in the early days of electric working.

The LCC Tramways introduced them on Saturdays, Sundays and Bank Holidays from November 1914. The MET and LUT introduced them in 1915, with a minimum fare of ½d covering the same distance as the 1d adult fare. The LCC began issuing children's tickets daily from July 1915, and it is possible that the LUT followed suit. The scale charged is not known, except that a ½d ticket has been seen carrying the same fare sections as the 1d adult ticket, so one can presume that half the existing adult fare was charged, including, at that time, fractions of a penny.

Children's fares were withdrawn when the ordinary fares were increased in December 1918 and were reinstated in July 1919, the scale being half the ordinary fare with a minimum of 1d, but this time fractions of a penny counted as 1d, the half-fares for the 3d and 5d adult being 2d and 3d respectively.

Appendix

Cheap Midday Fares

Generally, these fares as such were never introduced by the LUT. When LCC routes 2 and 4 were extended over LUT metals to Wimbledon Hill on May 2, 1922, the LCC scale of fares was extended through to Wimbledon. It included a 2d all-the-way cheap fare on Mondays–Fridays during the midday hours. Route 26, extended from Hammersmith to Kew Bridge on the same day, did not carry cheap midday fares west of Hammersmith Broadway, despite the fact that the LCC had become the owner of the tracks as far as Youngs Corner. This anomaly was not corrected until May 1, 1928.

When route 89 became a through-running service over LCC metals to Putney in November 1928, the LCC midday fare scale of 2d all the way and three sections 1d was applied between Acton Vale ('Askew Arms') and Putney. The track between the 'Askew Arms' and Hammersmith, although owned by the LCC from May 1922, had not been worked by the LCC until 89 was extended and became an LCC-LUT joint route.

Transfer Fares

From about the 1914–15 period, transfer fares were a feature of the LUT fare scale. The following examples are quoted from the tickets of the period:

1d Fares ($\frac{1}{2}d$ Child)

Park Road (Hampton Hill)–Teddington, Church Road, Change at Stanley Road Junction

Ham Boundary–Surbiton Station. Change at Eden Street Junction. Later, when 75 was curtailed at Kings Road, the transfer point was shifted to there

Summerstown–Tooting Terminus or Wimbledon Station. Change at Haydons Road Junction

$1\frac{1}{2}d$ Adult Fare only

Park Road (Hampton Hill)–Richmond Bridge. Change at Stanley Road Junction or York Street, Twickenham

Hampton Church–Teddington, Church Road. Change at Stanley Road Junction

Hampton Church–Kingston (Eden Street). Change at Hampton Court

Appendix

It is not known how long these transfers were in operation, as, with the fare increase authorized in 1918, geographical tickets disappeared in favour of the numbered variety, known to the conductors of that day as 'deaf and dumb' as they told people nothing unless they had an excellent knowledge of the numbers allocated to the various stages.

These tickets carry at the bottom a transfer section with cryptic code letters which will be discussed later under the section on tickets but which obviously indicate various fare stages.

From 1920 onwards, these code letters were removed in favour of a straightforward four-letter code which gives no idea of the transfer fares offered.

A further light is shed on LUT transfer facilities in the Kingston area in 1926, when LCC routes 2 and 4 were extended to Hampton Court on weekends. Geographical tickets were still being issued. The reverse side of the specimens seen reveals that transfers were still issued from Kingston (Eden Street) to the Karsino (Tagg's Island, Hampton) and Uxbridge Road, Hampton Hill, passengers changing cars at Hampton Court.

In 1924, when route 69 was withdrawn between Twickenham (York Street) and Richmond Bridge, 1d and 2d transfer fares were introduced from certain stages on the west side of Twickenham to the former tram stages at Twickenham (Crown Road) and Richmond Bridge, as well as through fares to Richmond ('Queens Head'), passengers changing at York Street to bus routes 27 and 33.

In 1935, two years after the LUT had been absorbed by the LPTB, the numerical tickets hitherto used at Hanwell Depot were superseded by a set of geographical tickets in the former LCC style, which revealed transfers on route 55 at Brentford Half Acre to Kew Bridge and Busch Corner, also at Hanwell Broadway to stages between Ealing Broadway and Hanwell Broadway, also to Southall Town Hall. A transfer stage from Hanwell Broadway to Greenford Road ('White Hart') was also revealed on the penny value, passengers changing at Hanwell (now called Southall) Bus Garage to the 18c bus.

From this it would seem that there were many transfer fares on the various fare tables which were not revealed to the student of this subject, owing to the use of numerical tickets and code letters for transfer fares.

It will be seen therefore that transfers had existed since their institution in 1915 and remained part of the fare scale until London Transport finally abolished all transfer fares on October 1, 1950.

Appendix

Cheap Return Fares

At Easter 1903 the LUT introduced circular trip fares covering the following journeys:

Hammersmith–Hampton Court–Hammersmith	1s
Richmond Bridge–Hampton Court–Richmond Bridge	8d

It is also known that at one time cars worked circular trips from Richmond Bridge to Richmond Bridge via Twickenham, Teddington, Hampton Wick, Hampton Court, Hampton Hill and Twickenham. Whether a similar circular operation from Hammersmith was carried out is not known, but it would have been possible to issue a circular trip ticket on cars from Hammersmith by changing at Twickenham. These tickets could be classified as an early form of return ticket.

The more orthodox style of return ticket was first introduced with the December 1920 fare increase on route 71 from Tooting to Hampton Court, the scale being as follows:

5d and 6d ordinary stages	9d	return
7d and 8d ordinary stages	1s	return
9d fare ordinary stages	1s 3d	return

At first these tickets were issued on Mondays to Fridays only, but *Tramway and Railway World* dated November 19, 1921, stated that these return fares were so successful that they were to be issued on Sundays, and, in addition, the following return fares would be issued on route 69:

Kingston (Eden Street)–Twickenham Junction	5d
Kingston (Eden Street)–Richmond Bridge	6d

These fares must have been withdrawn not long after as there is no mention of them in the Map and Guide for 1922.

A cheap return fare of 9d was charged between Wimbledon Station and Hampton Court on route 71 in 1928. This fare was available on weekdays only and was also available on LCC routes 2 and 4 when running to Hampton Court on Saturdays.

On Wednesday January 1, 1930, a revised scale of return fares was introduced on route 71 as follows:

3d return for any 2d fare stage
5d return for any 3d fare stage

Above: A modern view of the former Chiswick power station. Note the stone plaque over the doorway with its female figures and the initials LUET. (*London Transport*). *Left:* The curved staircase and gallery in the former Chiswick power station. (*London Transport*). *Below:* Wellington Road entrance to Fulwell Depot in 1933, with trolleybuses (*London Transport*)

Top: Trolleybus on Hampton Court–Wimbledon service in Wimbledon Hill Road. (*London Transport*)
Centre: Trolleybus No. 43, in original condition, without large central headlamp, turns at Hampton Court in September 1931. (*London Transport*)
Bottom: 74-seater centre-entrance experimental trolleybus, March 1933. *London Transport*)

Appendix

6d return for any 4d fare stage
7d return for any 5d fare stage
8d return Wimbledon–Hampton Court

The introduction of this scale caused the reduction of the throughout return fare introduced in 1928 from 9d to 8d. Once again these fares were also applied on routes 2 and 4 when running to Hampton Court.

At the same time, cheap return fares were introduced on 67 between Hammersmith and Hampton Court to the following scale:

8d return for any 6d single fare
9d return for any 7d single fare
10d return for any 8d single fare
11d return for any 9d single fare
1s return Hammersmith–Hampton Court

These tickets were issued daily at any time on route 71 but those on 67 were issued from 10 a.m. on weekdays, all day on Sundays but not on public holidays.

In 1931 return fares on a restricted scale were put into operation on route 7 (Shepherds Bush–Uxbridge), the charges being:

1s return for 7d and 8d single fares
1s 3d return for 9d and 10d single fares
1s 6d return Shepherds Bush–Uxbridge

The times and days of issue of the above fares were as for route 67.

All these return fares, suitably increased with the fare increase on February 9, 1947, were available until the withdrawal of all return fares on trams and trolleybuses on October 1, 1950.

Through Booking Fares with the Underground Railways
It can be said that Sir Clifton Robinson was the instigator of this type of facility, as through booking fares between points on the London United system and stations on the Piccadilly and District Railways commenced with the opening of the Piccadilly tube on December 15, 1906.

They were available from Hampton Court, Hounslow and Uxbridge, also intermediate fare stages on the LUT, to stations as far as Whitechapel (District Railway) and Finsbury Park (then the northern terminus of the Piccadilly).

The fares charged were a compilation of the tram fare from any fare

Appendix

stage on the above three routes, plus the railway fare from Hammersmith to the station in question on the two Underground lines.

For example, a passenger travelling from Twickenham to Finsbury Park would pay 8d, consisting of 4d tram fare to Hammersmith, plus 4d tube fare to Finsbury Park.

A careful census must have been taken of the number of these tickets issued, as revised arrangements were brought into force on July 1, 1912.

Whilst the facilities were still available as far as Finsbury Park and Whitechapel on the railways, the tramway availability was reduced to the following points:

Hampton Court and Hounslow Routes
Turnham Green Church, Gunnersbury Station, Brentford Half Acre, Hampton Court

Hanwell Route
Acton (Birch Grove), Ealing Broadway, Hanwell Broadway

By 1913 the bookings to Ealing and Hanwell were available via Ealing Common Station as well as Hammersmith.

On February 1, 1914, further facilities were provided to the Central London Railway via Shepherds Bush Station from Acton (Horn Lane), Acton Vale ('Askew Arms') and Youngs Corner.

On the same date, through bookings were added via Boston Manor Station to Brentford Half Acre and Hanwell Broadway via LUT route 55.

How these through bookings were affected by the 1918 and 1920 fare increases is not known, but the Tramways Map and Guide for Summer 1921 only advertises through bookings via Boston Manor Station.

Tramway and Railway World in its issue of March 11, 1922, mentions that through bookings to Underground stations from the LUT had been initiated but no date of commencement is given.

From the summer edition of the Map and Guide it would seem that bookings existed as follows:

Tramway Point	*Railway Exchange Point*
Youngs Corner	via Hammersmith or Shepherds Bush
Turnham Green Church	via Hammersmith or Shepherds Bush
Kew Bridge	via Hammersmith or Shepherds Bush (also via Chiswick Park)

Appendix

Brentford GW Station	via Hammersmith, Shepherds Bush or Chiswick Park
'Askew Arms'	via Hammersmith or Shepherds Bush
Acton (Horn Lane)	via Hammersmith or Shepherds Bush
Hanwell Broadway	via Ealing Common, Shepherds Bush or Boston Manor
Brentford Half Acre	via Boston Manor

These tickets were issued as far as Liverpool Street on the Central London Railway, but only as far as Piccadilly Circus and Charing Cross on the Piccadilly and District lines, although they were now available to stations on the Hampstead and Bakerloo railways.

At the same time, rates were introduced for tram-rail season tickets via Boston Manor to Brentford and Hanwell, and via Ealing Broadway to West Ealing (Melbourne Avenue).

Later in 1923 season ticket rates were introduced to tramway points via Hammersmith and Shepherds Bush station but not via Chiswick Park. Those via Hammersmith to Starch Green ('Seven Stars') were available via route 89. Those via Shepherds Bush were available to Starch Green or Youngs Corner on route 57, and 'Askew Arms' and Horn Lane, Acton, on routes 7, 83 and 87.

By the winter of 1925, the issue of these TOT seasons (as these tickets had become known) via Ealing Broadway had been extended to fare stages on route 7 as far as Hayes ('Adam and Eve') and were available via either Ealing Broadway or Ealing Common stations.

All these season ticket rates were for periods of one month or quarterly, but weekly seasons were introduced in April 1928.

These special fare facilities remained constant throughout the remainder of the life of the LUT but a census of the use of these fares was taken in 1935 and on October 1 many were withdrawn as they were little used, and only the more popular fares were retained, both for ordinary singles and season tickets.

The single journey through fares which remained after 1935 were withdrawn in October 1942, the reason given being that the female staff recruited during the war could not cope with their issue under wartime conditions.

The TOT special ticket rates remain in existence to the present day, although the issue of such tickets is restricted to existing users and no new intending passengers can use the facility.

Appendix

Through Bookings with LGOC Buses

After the Kew Road horse tramway was given up, through bookings were introduced with LGOC route 27 as follows: Richmond (The Quadrant)–Stanley Road Junction (1½d); Teddington (Church Road) (2d); Hampton Wick (Wick Road) (2½d); and Kingston (Eden Street) (3d). Passengers changed from bus to tram at Richmond Bridge tram terminus. These fares were withdrawn on and from October 1, 1917.

When the LGOC introduced its first country bus service from Hounslow ('Bell') to Windsor Castle in 1912, through bookings were brought into force from points on tram routes 57 and 61 to Windsor, passengers changing to bus at Hounslow ('Bell'). These were probably withdrawn when daily operation of the bus route ceased in 1915 for the duration of the war. The facility was not reintroduced when bus services from the country to Hounslow commenced operating once again, first on Sundays only, then finally daily.

As stated, when tram 69 was withdrawn between Richmond Bridge and Twickenham in 1924, through fares were introduced to Richmond via buses 27 and 33 and remained a part of the fare table until the end of all transfer fares on October 1, 1950.

Fare Collection

For the whole of its life the LUT relied mainly on the Bell Punch ticket system as the most reliable method of fare collection. On the one-man cars on the Richmond Park Gates–Tolworth and Brentford–Hanwell routes 'Automaticket' machines were used. They issued a coloured ticket about the size of the present-day cinema ticket. As there was no space to indicate the stage of boarding or alighting, dishonest passengers must have had a good time!

In the latter part of 1932, the conductors from Hounslow and Fulwell Depots lost their Bell Punches and coloured tickets and the TIM machines were used, even for the issue of through booking tickets to the Underground. Bell Punch tickets were then only seen on tram routes 57 and 67 and trolleybus routes 1, 2, 3 and 4 when the TIM machines broke down, although for a long time from July 1940 these machines were not capable of issuing 1d child's and 2½d workmen's tickets, and the Bell Punch version was used, the tickets being punched in the small cancelling machine provided on the holding strap.

Tickets – Horse Era

The Shepherds Bush–Youngs Corner route required only 1d ticket,

Appendix

which was coloured white. It carried the company's name up the centre of the ticket and the usual regulation that it must be retained intact and shown or given up on demand. It was of the fully geographical type and the fare was set out as follows:

Left-hand side: Uxbridge Road Station to
                  Youngs Corner, Chiswick
Right-hand side: Youngs Corner, Chiswick
                  to Uxbridge Road Station   (type 1)

(The type numbers given at the end of each description refer to the position of the ticket in question in the pages of photographic reproductions)

When this route was extended to Kew Bridge on Sundays, thus bringing it into conflict with the route from Hammersmith to Kew Bridge, it was felt that the tickets on Shepherds Bush cars should be distinguished from those on the Hammersmith route. Accordingly, the 1d ticket became white with a vertical red stripe (Hammersmith ticket plain white), with the 2d ticket for the whole distance becoming orange (Hammersmith ticket blue).

On the introduction of electric traction these colours were retained and the orange colour was also used for the 2d value on the Southall route.

Tickets – Electric Era
With the inauguration of the Hounslow and Hampton Court routes a new colour scheme was adopted: 1d white, 2d blue, 3d pink, 4d green, 5d flame and 6d blue. As the routes were now getting longer, it was no longer possible to use the full geographical layout on the lower values and these were set up in what is known to ticket enthusiasts as 'menu style' (type 2).

Tickets from 3d upwards retained the fully geographical layout, with the exception that the fares were only set up sideways on the tickets for the full extent of the route (types 3 and 4).

Despite the set-up of the colour scheme introduced on the Hampton Court and Hounslow routes, when the Southall route was extended to Uxbridge a new colour scheme was introduced for its tickets, as follows: 1d white, 2d blue, 3d yellow, 4d green, 5d cedar (type 5).

Workmen's singles were white with either a red or green overprint diagonally from left to right as follows: 'Workmen's Ticket'. The work-

Appendix

men's returns cannot be described as they were withdrawn on the return journey and an exchange ticket issued. The only other overprint used at this time was a large 'K' in red on the tickets used on the local Kingston routes, although the exchange tickets for workmen's returns had this letter superimposed in green (type 6).

To assist the clerks employed at depots replenishing the ticket boxes, a code indicating the route was printed either in the centre column at the top or in the bottom left-hand corner. These codes were as follows:

HHC	Hammersmith–Hampton Court
SHC	Shepherds Bush–Hampton Court
SH	Shepherds Bush–Hounslow
HH	Hammersmith–Hounslow
HK	Hammersmith–Kew Bridge
SK	Shepherds Bush–Kew Bridge
SS	Shepherds Bush–Southall
THC	Tooting–Hampton Court
RK	Richmond–Kew
U	Uxbridge–Hammersmith or Shepherds Bush

The fare revision of 1910 brought a revised colour scheme, as, for the first time, odd value fares had to be catered for. At the same time, a revised set-up for the printed matter on the tickets themselves was introduced. On this series of tickets, the fare stages were set up horizontally on each side of a central column, which carried the fare value in a box at the top (horizontal) and the name of the company and revised conditions of issue set vertically up the column. These conditions read 'Passengers must not break their journey. To be punched in the section to which the passenger is entitled to travel'. The fully geographical aspect disappeared. The left-hand column carried the fare stages from the outer terminal in descending geographical order until the stage covering the last penny fare to the London terminal was reached. The right-hand column, commencing with the stage at which the first penny fare from the outer terminal ended, carried the fare stages in descending geographical order to the London terminus at the bottom. On journeys outward from London, the left-hand side of the ticket was punched, with the right-hand side being used on inward journeys (type 7). The revised colour scheme was as follows:

Even fares *Odd fares*
1d white 1½d white with mauve edges

Appendix

2d buff 2½d buff with brown edges
3d pink 3½d buff with orange centre stripe
4d green 4½d green with brown edges
5d flame 5½d orange with brown edges
6d blue (this was the highest fare value at this time)

It would seem that separate tickets were used this time for transfer fares. As these were collected on the second car and a standard buff exchange ticket was issued in their place, no specimens have come to light.

About 1914–15 a new style of ticket was adopted which was contemporary with the LGOC bus ticket where the fare stages were set up centrally in wide boxes with the name of the company and the sentence 'This ticket to be shown on demand'. The fare value was set up vertically in the left-hand gutter, whilst the right-hand gutter carried the instruction to the conductor that the ticket had to be punched in the section to which the passenger was entitled to travel.

On the LGOC tickets the service number was carried in an elongated box at the top of the ticket, but, despite the fact that all LUT routes had been numbered by this time, the numbers were not used on the tickets. Instead, the route was indicated in this box as follows: Tooting, Hounslow, Uxbridge, Richmond Bridge, Kew Bridge.

Some types, which could be used on several routes, were headed as follows: H, S B and H C (Hammersmith, Shepherds Bush and Hampton Court), also a very full title reading 'Tooting & Kingston Cross Town Routes' has been seen, indicating that a separate set for the Kingston locals had now disappeared.

With the introduction of this style of ticket, a few colour changes took place, the 1½d value becoming lilac instead of striped. The 4d value has been seen in brown as well as the usual green. The 5d value appeared in grey in addition to the more normal flame, possibly because of the difficulty in obtaining suitably coloured paper during the war years. The ½d child fare introduced in 1915 was catered for with a salmon-coloured ticket.

Although the fare value was printed in the left-hand gutter at the top, value overprints in either red or green were superimposed on the printed matter, possibly because of the use of varying coloured papers (type 8).

At this time the transfer fares started to appear on the single tickets

Appendix

where they applied, and were printed below the ordinary fares, showing the full extent of the fare and the transfer point (type 9).

Exchange tickets were still issued but lost their fully geographical style, being set up similarly to the ordinary single tickets, and were issued in exchange for workmen's returns and transfers. At first, they retained their buff colouring but later in the 1914–18 war went to white paper, which at times deteriorated to a very dirty off-white shade (type 10).

These beautifully printed geographical tickets remained in issue until the advent of the fare increases of December 1918.

This fare increase is notable for three things:

(a) the disappearance of geographical tickets, fare stage numbers from 1 to 24 being used to denote the stages
(b) the end of odd value fares such as 1½d, 2½d, 3½d, etc.
(c) the issue of adult fares above sixpence for the first time

Colour changes were also made, the 4d ticket returning to its original green; the 1910 colours were retained for the other tickets in the 1d–6d range, whilst the higher values were allotted colours as follows: 7d brown; 8d lilac; 9d yellow; 10d grey.

From the typesetting angle, there appears to have been two sets of tickets produced for this fare change. One carried the heading 'Uxbridge' at the top, fare stage numbers from 1 to 24 and was for use at Acton and Hanwell depots which supplied the cars for routes 7, 83, 87 and 89. This set has no transfer sections at the bottom, no fare values in the text, and the same block could be used for all tickets, the fare value being indicated by the colour and overprint (type 11).

The other set, which was headed 'London' for some reason on the values from 3d upwards, carried stage numbers from 1 to 20. The 1d and 2d values only carried a transfer section at the bottom with letters in code which obviously refer to fare points indicating the extent of transfer fares.

The 1d value carries the following letters: P, S, and SS in the left-hand column, with T, TW, and HK on the right-hand side. As the letters S and TW are opposite each other this would seem to allude to transfers between Summerstown (S) and Tooting or Wimbledon Station (TW). The letters SS are set opposite HK; SS decodes itself as Surbiton Station, whilst HK could indicate Hampton (The Karsino), at that time a stage on route 67. The letters P and T have defied all efforts to decode

Appendix

them, although T could refer to Tooting, Tolworth or even Twickenham (type 12).

There are two tickets in the 2d value in this twenty-stage series, one carrying the letters HC and ES only in the transfer section, indicating a transfer fare from Hampton Church to Eden Street (Kingston). The other ticket carries the letters H and TE in the transfer boxes, letters which have defied decoding (types 13 and 14).

No children's tickets have been seen for this period. As the associated MET had withdrawn this facility with its 1918 fare increase, it would seem that the LUT had done likewise.

With the 1920 fare changes, a standard set of numerical stage tickets appeared, bearing numbers from 1 to 30. New tickets for $1\frac{1}{2}$d (minimum adult fare) and 1d (minimum child fare) appeared. The $1\frac{1}{2}$d took the white colour of the former 1d ticket, whilst the 1d child was printed on a very dark all-over sage green paper.

Three basic blocks were used for the printed matter. One was $3\frac{1}{8}$ inches deep and, carrying a transfer section at the base lettered A–D, was used for 1d child, $1\frac{1}{2}$d and 2d values. A block of the same size but minus the transfer sections was used for 3d and 4d values, and a block with the stage numbers set closer together, making a ticket $2\frac{1}{2}$ inches long, was used for 5d to 10d values (types 15, 16 and 17).

Workmen's tickets now consisted of returns only which were printed in the same colours as singles but distinguished from them by having the fare value incorporated in the text, and a large cross overprint on the printed matter (type 18).

At first these tickets were withdrawn on the return journey and a rather wasteful series of exchange tickets, all coloured salmon, but covering values from 2d to 10d, was issued in exchange (type 19).

Much valuable revenue was lost at this time by the withdrawal of commercial advertising from the reverse side of the ticket, as the conditions of issue were printed on this side (type 20).

When the 1d fare returned for one half-mile fare stage in 1921, the 1d ticket took its original white colouring, with the $1\frac{1}{2}$d value for two sections becoming white with a red bar overprinted on both edges (type 21).

For some unknown reason, the universal set of tickets did not meet the requirements of the Traffic Department, and although the basic format was retained, a code letter indicating the depot of issue was added in the centre column (type 22).

Appendix

The codes used were: A – Acton; C – Chiswick; F – Fulwell; HL – Hanwell and HW – Hounslow.

The last major fare change came into force on January 1, 1923, the only alteration affecting the tickets being the withdrawal of the 1½d adult fare except on route 81 (Haydons Road Junction–Summerstown). This route was too long for a 1d stage and too short for a 2d stage, so the throughout fare was reduced from 2d to 1½d.

The colour scheme was to remain static for the rest of the LUT's life. The colours were:

1d white	5d flame	10d grey
1½d white (red bars)	6d blue	1d (child) sage green
2d buff	7d brown	Exchange: salmon for workmen's tickets only
3d pink	8d lilac	
4d green	9d yellow	rail exchange: salmon with green centre bar

Overprints of the fare value were used on all ordinary single tickets. These were of the skeleton type and printed in red, although in the 1930s the overprint on 3d and 5d values was printed in dark green (type 23).

Workmen's tickets, which were returns, were of the same colours as the single ticket of the value in question, the range being from 2d to 10d. They were distinguished from the singles by an overprinted cross in red, but once again the 3d and 5d values were issued with a green cross overprinted from about 1930 onwards (type 24).

The format of all tickets was changed once again in 1926, when a return was made to the practice of displaying commercial advertising on the reverse side of all tickets, with the exception of road-rail through booking tickets.

Opportunity was taken to reset the stage numbers in the centre of the ticket with the company's name, depot code and the words 'Issued subject to Company's by-laws and available to point indicated by number in section punched' in the left-hand gutter. In the right-hand gutter, two warning sentences were set, reading 'To be shown on demand. Break of journey not permitted' (type 25).

The lower value tickets were reduced in size to 2½ inches, this being done by setting the transfer codes in line across the base of the ticket. The wording in the gutters was also slightly different. In the left-hand gutter, the passenger was informed that the ticket was available to point indicated by the number or letter in the section punched, whilst the

Appendix

right-hand gutter said that break of journey was not permitted unless the ticket was punched in the lettered section (type 26).

The depot codes were changed, and the tickets for each depot carried the number of stages required for use on the routes from that depot, as follows:

Acton	Depot name in full. Twenty-six stages on ticket
Chiswick	This depot had passed to the LCC tramways by this time
Fulwell	Code FLWL. Twenty-four stages on ticket, although a twenty-six-stage series seems to have been issued for a short time
Hanwell	Code HNLL. Twenty-six stages on ticket
Hounslow	Code HLOW. Twenty-four stages on ticket

Also with this change of format, special cancelling punches for the cancellation of return tickets were introduced, and the wording on workmen's return tickets had the instruction covering the return journey altered to read 'To be handed to the conductor for cancellation'.

About 1928 the tickets used at Fulwell Depot were divided into two sets:

(1) FLWL Stages 1–24.	For use on route 67. Values 1d–10d and 1d child's. Workmens returns 2d–10d
(2) FLWL Stages 8–31.	For use in Kingston area on routes 69, 71, 73, 77 and 81. Values 1d–6d. 1d child's and 1½d (route 81 only), also 9d ordinary return (route 71) (type 27). Workmens returns 2d–6d only

The 8–31 set did not last very long, as on January 1, 1930, a revision of stages on route 71 between Norbiton Church and Malden Fountain was brought into operation. Gloucester Road ceased to be a fare stage and Norbiton Church and Kingston Sports Ground were included, thus adding an extra fare stage. This affected the stage numbering from Wellington Crescent (Malden) to Wimbledon Station, making the latter point stage 32 instead of 31. Pending the issue of fresh tickets, conductors were instructed to punch existing tickets in stage No. 8, which would represent Wimbledon Station.

The 9d ordinary return fare issued on weekdays only on route 71 was

Appendix

withdrawn at this time and cheap return fares at 3d, 5d, 7d, and 8d took its place.

These tickets were of the same colours as the corresponding single values, but the stage numbers were set in ladders on the left- and right-hand edges of the ticket, whilst the centre carried the conditions of issue. To distinguish them from workman returns a small fare value in italicized skeleton type was printed thereon. Whilst the lower values carried stage numbers (type 28), the 8d which covered the full journey was geographical (type 29).

To return to single fares for a moment – two new values were added to the set used at Hanwell Depot, although these appeared only on Sundays, Bank Holidays when traffic was heavy and for special traffic checks. These were 11d single – deep red (blue overprint) and 1s single – deep magenta (red overprint).

At the same time as the return fares came into operation on route 71, 8d, 9d, 10d, 11d and 1s return tickets appeared on route 67 for fares between Hammersmith and Hampton Court. The 8d to 10d values took the same colours as single tickets with fare stage numbers in the range 2 to 21. The 11d value has not been seen but the 1s value was white with dark blue bars covering the fare stage numbers, and which would appear on the clipping left in the ticket punch.

In 1931 cheap day returns on a limited scale were introduced on route 7 to Uxbridge. Three tickets were required, 1s, 1s 3d and 1s 6d, carrying fare stage numbers in the range 2–26. The 1s value was the same colour as that used for Hampton Court, whilst the 1s 3d was white with red bars, and the 1s 6d was white with green bars and geographical stages (types 30 and 31).

The inauguration of the trolleybus system in the Kingston area in 1931 affected the set of tickets used from Fulwell on the trams in this part of the system.

The set carrying stage numbers from 8 to 32 disappeared and was replaced by a set with stage numbers from 13 to 32 with transfer sections lettered from A to D on all values. Conditions of issue, colours and style were the same as for the previous tramway tickets and the values ranged from 1d to 6d (type 32).

As no alterations were made to the return fares or the stage numbers when route 71 became trolleybus 4, the cheap return tickets remained unaltered.

After a few months, the single tickets and workmen's returns were

Appendix

replaced by a new style of ticket. The size was increased from 2¼ to 2½ inches and the stage numbers were printed in a very much bolder type, whilst transfer sections from A to H appeared on all single values and the workmen's returns as well. In the absence of a fare table for that time, it must be presumed that there were many transfers added in the Kingston area and that the facility had been extended to workmen's return fares (types 33 and 34).

In November 1928, through running over LCC metals came about with the weekday extension of route 89 from Hammersmith to Putney.

There was an immediate problem in that the major portion of this route was in the LCC area (Putney–'Askew Arms') and the LCC was naturally anxious to extend its maximum 2d midday, 5d ordinary, 6d workman and 8d return fares from London to 'Askew Arms'.

Hitherto, as route 89 had always terminated at Hammersmith, this point had been stage No. 1. The LCC rightly pointed out that Hammersmith was no longer the terminus for this route and suggested that the stages should be numbered to conform with its services 26, 28 and 30, on which Fawe Park Road was No. 14. The LUT agreed and the stages were numbered from 14 to 25, which could easily be accommodated by the Acton Depot ticket, which already carried numbers from 1 to 26.

For ordinary single fare journeys up to 4d and for 2d and 4d journeys (workman's) the LUT used its own tickets.

To cover all journeys with a transfer facility in the LCC area, LCC tickets were supplied to the LUT for use. These values were as follows:

5d single and 2d child
cheap midday fares – 2d adult and 1d child
ordinary returns – 5d, 6d, and 8d
workman returns – 6d (types 39 and 40)

The through fares to the Underground introduced in December 1906 were a new style of facility. Some method had to be introduced whereby each operator could be sure that he was receiving the portion of fare to which he was entitled, and for the tram-to-rail ticket The Bell Punch Company produced one of its famous 'Duplex' type tickets.

Instead of being printed on the normal stiff card, these Duplex tickets were printed on thin paper twice the width of a normal ticket and were perforated down the centre. They were folded to normal ticket size and were the same thickness as an ordinary ticket when placed in the punch. Upon issue, they were punched twice, once in the appropriate fare section

Appendix

on the tramway route, which indicated to the Audit Department how much was due to the tramway company, and once in the railway section.

Different coloured paper was used for each of the routes issuing through booking tickets, blue being used for the Hampton Court route, white for Hounslow and red for Uxbridge. The tickets were 5 inches long and the portion issued to the passenger was overprinted with a large skeleton numeral indicating the month of issue. For instance, a large 7 indicated the month of July.

The ticket issued from the railway stations was a standard Edmonson type, and in the early days was a whole ticket, reading, for example: Piccadilly Circus to Hammersmith by Piccadilly Railway, thence by London United Tram to Kew Bridge. To be given up on Tram for an Exchange Ticket'. Later, the tickets were of the style used for returns normally, with the left-hand half available on the railway and handed to the collector at the exchange station, whilst the right-hand portion was handed to the tram conductor who issued an exchange ticket in its place. This second arrangement lasted throughout the whole existence of the tramway companies.

The exchange tickets issued were also of Duplex type until they were withdrawn during the First World War, and were very wide tickets on pink paper at first. Later they became normal width and were printed on green paper (type 35).

When these bookings were reinstated in 1922, the tram-to-rail singles were standard type Bell Punch tickets printed in the same colours as the ordinary single tickets (type 36), whilst the exchange became a standard salmon-coloured ticket with green centre stripe (type 37), although after workman's exchange tickets were withdrawn, a new short-style ticket appeared in salmon only, and carrying both fare stage numbers and names in the centre of the tickets (type 38).

When LCC route 26 was extended into the LUT area, the LCC used LUT through booking tickets for fares via Chiswick Park and Hammersmith, but upon the introduction of route 89 as a through-running route in which the LCC participated, this authority produced a set of through booking tickets for both 26 and 89 for use on its own cars and coloured in its famous striped style.

The advent of the TIM machine at Hounslow and Fulwell depots brought the use of all types of printed ticket to an end from those depots. A list of fare stage numbers was issued to Underground ticket collectors and they accepted TIM tickets bearing the appropriate stage number for

Appendix

the stage from which a through booking applied. The other depots continued to use the Bell Punch printed tickets, which remained in LUT colours until 1935, when the LPTB started printing all tram and trolleybus tickets, and all through booking tickets were printed in the former LCC striped pattern. The colour chart was as follows:

LUT Colours	*LPTB Colours (Ex LCC)*
3d pink	3d blue/white centre stripe to 1937
4d green	pink/white centre stripe from 1937
5d flame	4d green/white centre stripe
6d blue	5d brown/white centre stripe
7d brown	6d primrose/white centre stripe
8d lilac	7d rose pink/white centre stripe
9d yellow	8d grey/white centre stripe
	9d blue-grey/white centre stripe

The striped tickets remained in issue until all through booking single fares were withdrawn in October 1942.

The railway season tickets were in various colours, mainly distinguishing the period of issue (monthly or quarterly) and were fully geographical, reading as follows:

Piccadilly Circus to Hammersmith thence by L.G.O. Bus or L.U. Tram to Turnham Green Church

The above is, of course, just an example of the printed matter. As a guide to the tram conductors, two or three red or green stripes were printed over the top half of the ticket, but these were later superseded by a number in italicized skeleton type overprinted on the printed matter in red and which indicated the number of fare stages to which the passenger was entitled to travel.

The authors will be grateful to receive further information on matters which have not been fully covered, both on fares and tickets.

TYPES OF VEHICLES

ELECTRIC TRAMCARS

Type (after 1912)	Nos.	Year built	Type	Body	Trucks	Motors	Controllers	
Z	1-100	1901	double deck open top short canopy	Hurst, Nelson Motherwell	Peckham 14 D 2 max. traction swing bolster	BTH GE58 (2) 25 h.p.	BTH BT8	52 ran in 1909 with Barber 6-wheel truck
X	101-150	1902	double deck open top short canopy	Geo. Milnes Birkenhead	McGuire max. traction swing bolster	Westinghouse 49B (2) 25 h.p.	Westinghouse 90	Little used after 1910 and stored 1914. 108, 118, 125, 137, 149, 150, bought in 1919 by Blackpool for £1,438 10s. (one account states for £425 each and without motors, controllers and armatures) 107 sold to MET in 1921. 101, 102, 103, 104, 105, 106, 109-116, 119-124, 126-136, 139-147, sold to LCC and re-purchased. 141 and 142 rebuilt as 'flood cars'. 148 finally Hanwell works car
W	151-211 237-300	1902	double deck open top short canopy (except Nos. 175 and 275, see remarks) double deck open top short canopy	Milnes	Brill 22E max. traction side bearing	Westinghouse 49B	Westinghouse 90	165, 173, 183, 200, 240, 254, 259 reconditioned 1927-8 and fitted with Metrovick 104 motors and magnetic brakes 157, 161, 211, 243 rebuilt 1928 and, with 261 (type U) became type WT on receiving type T top deck but with deeper end panels
	212-235	1902		British Electric Car Co.		BTH GE58	Westinghouse 90	175 and 275 built as, or altered to, single-deck 'social' saloons (see also S2) 178 rebuilt 1924 as single-deck, one-man car (see also S2) 187, 192, 221, 252 leased in 1915 to Erith, which bought them in 1919 at £570 each 226, 227, 228, 229, 230, 232 sold in 1919 to Walthamstow at £570 each In 1928, 155, 199, 288 received top covers of old type, GEC WT28 motors, and BTH B49 controllers, and became type U2
WT (see remarks)								Reconditioned in 1926-7 with high-power motors and magnetic brakes
(U ex W)			The following 50 cars received top covers with short canopy in 1910-11: 151, 152, 154, 158, 171, 204, 205, 210, 212, 224, 235, 236, 247, 250, 251, 255, 256, 261, 262, 265, 266, 267, 268, 269, 270, 271, 272, 273, 274, 276, 277, 278, 279, 281, 282, 283, 284, 285, 286, 287, 290, 291, 292, 293, 294, 295, 296, 297, 299, 300						261 became type WT in 1928 on fitting of type T top deck 247 of type U reconditioned about 1928 with Metrovick 104 motors and Westinghouse 90M controllers. It had type X lower deck and type U upper deck, it was classified type XU In 1928 type W cars 155, 199, 288 fitted with top covers of old type, GEC WT28 motors and BTH B49 controllers, and reclassified type U2. 288 received plain arch roof to lower deck and transverse seating in saloon In 1931 ten cars exchanged for ten SMET open-top cars. Cars finally on loan to SMET were: 267, 268, 269, 271, 272, 276, 278, 290, 293, 299
XU (see remarks)								
U2 (see remarks)								
T	301-340	1906	double deck top covered balcony	United Electric Car Co.	Brill 22E max. traction side bearing	Westinghouse 80 (30 cars) Westinghouse 200 (10 cars) later, (2) 40 h.p. Metrovick 60 h.p.	Westinghouse 90, later 90M	310-323 sold to LCC and re-purchased In 1925, all class received high-power motors of GEC WT 28 type and magnetic brakes, and transverse seats in saloon. 307 fitted with upholstered seats on top deck About 1930, 301 fitted with type WT deep end-panels to top deck In 1932, between 13 and 20 cars fitted with windscreens When type UCC cars (see below) delivered in 1931, about 12 type T cars retained on Uxbridge route. Remainder to routes 55, 57 and 89. Finally transferred to MET

Class	Numbers	Year	Body	Truck	Electrical equipment	Notes
Y (ex Z)			The following 52 cars received top covers with short canopy in 1910–11: 1, 2, 9, 10, 12, 13, 14, 15, 18, 22, 24, 25, 27, 28, 29, 30, 31, 32, 35, 36, 37, 38, 39, 43, 47, 48, 49, 56, 59, 61, 65, 67, 68, 69, 70, 71, 72, 74, 75, 76, 78, 79, 82, 85, 87, 91, 92, 93, 94, 95, 98, 99				
S (S1)	341	1905 (1922–4)	Single-deck one-man operated	Brush, reconstructed 1922 by MET, rebuilt 1924 by LUT	Brush-Warner radial, then Brill 22B max. traction	BTH GE67 then BTH GE58	Originally MET type E No. 132, a single-deck, 6-windowed car on Brush-Warner radial truck. Rebuilt by MET as 5-windowed car. Again rebuilt by LUT in 1924 as 6-windowed car on original underframe, mounted on Brill bogies and vestibuled (first fully vestibuled car in London). Withdrawn 1928
S2	342–344	1902 (1924)	Single deck one-man operated	Milnes, reconstructed 1924 by LUT	Brill 22B max. traction	BTH GE58	Reconstructed in 1924 from type W Nos. 175, 178, 275 and vestibuled. Withdrawn 1928
'Poppy'	350	1928	double deck top covered	LGOC Chiswick Works	Brush M & G max. traction swing bolster with Hoffmann roller-bearing axleboxes	BTH B18	Retrucked at Hendon Works with plain axlebox bearings
UCC ('Feltham')	351–396	1931	double deck top covered all enclosed	Union Construction & Finance Co.	English Electric max. traction swing bolster (No. 396 had special flexible drive and equal wheel trucks	GEC W729 (No. 396—DK131)	On delivery these cars took over almost all workings on the Shepherds Bush–Uxbridge route and Sunday workings on the Brentford–Hanwell route. They continued on these services after the formation of London Transport until displaced by trolleybuses in 1936. They were then transferred to former LCC routes in South London, from which they were in turn displaced by diesel buses in 1951. All but six, scrapped between 1947 and 1950, were sold to Leeds City Transport, with 52 former MET cars of the same type, and ran in the service of that undertaking until 1959.

001—4-wheel sand van
002, 003—rail polisher carrying water tank on 4-wheel truck
004—4-wheel ticket van (built BE Car Co, 1903)
005—stores van (bogie car, trucks from type W car)
006—rail polisher. Double-deck open top, on 4-wheel truck. Ex-Croydon Corporation No. 4, one of 12 cars bought from Croydon by SMET.

TROLLEYBUSES

Class	Numbers	Year	Body	Chassis	Notes
A1, A2 ('Diddlers')	1–35, 36–60		double deck top covered 6-wheel AEC 663T chassis half-cab layout	Union Construction	A1 had EE electrical equipment, with 80 h.p. DK 130A motor; A2 had BTH electrical equipment with 82 h.p. BTH 110 DL motor. Half-cab design. Nos. 1–59 seated 24 in lower and 32 in upper saloon. No. 60 seated 27 in lower and 29 in upper saloon. No. 1 is preserved at the Museum of British Transport, Clapham. (The origin of the nickname 'Diddlers' is obscure. One suggestion is that the front-end appearance 'diddled' passengers into believing that a type LT motorbus was approaching. Another is that, as the vehicles were not fitted with batteries, it was necessary to 'diddle about' to move one that had stopped with its trolley wheels on a 'dead' point, such as a crossover in the overhead.)
X1	61	1933	double deck top covered centre-entrance full frontal layout	LGOC Chiswick AEC 691T chassis	EE 404 motor, 95 h.p. Centre entrance. 74 seats

Appendix III

The Fleet

The LUT took over 251 horses and 33 horse cars from the West Metropolitan. By June 1897 the fleet had grown to 366 horses and 49 cars, and by June 1901 to 482 horses and 59 cars, and 100 electric cars.

Between the end of 1903 and that of 1911, the last full year of horse traction on the Kew Road line, the total of horses and horse cars declined from 83 to 61 and from 10 to 9 respectively, while that of electric cars rose from 300 to 340. Sales, scrapping and withdrawal from service for storage had reduced the fleet to 250 electric cars by the end of 1922 and to 189 by the end of 1930.

The fleet total at the end of 1931 was 150 cars and 60 trolleybuses. The company handed over 149 cars and 61 trolleybuses to the London Passenger Transport Board in 1933.

There has been much controversy concerning the colour of LUT cars. The following table summarizes the information available on the first five years of electric operation:

Car Nos	colour	routes allocated
1–100 (later type Z)	scarlet and white	Hammersmith/Shepherds Bush–Kew Bridge/Hounslow
101–150 (later type X)	white	Shepherds Bush/Hammersmith–Ealing–Southall/Uxbridge
151–300 (later type W)	royal blue	Hammersmith/Shepherds Bush–Hampton Court, and Richmond Bridge section
301–340 (later type T)	scarlet	Originally Kingston area then transferred to Uxbridge routes

Appendix

It has been suggested that the one colour one route idea, following horsebus practice, allowed easy identification in an era when many passengers were still illiterate. One authority states that the white livery was discarded about 1905 as being costly to maintain and that for a time yellow replaced it. Another source has it that some cars sent to replace the 301–340 batch on the Kingston lines were repainted yellow to console Kingstonians for the loss of their more modern vehicles! There is also a suggestion that some cars were repainted brown for a time.

From about 1920 the Underground Group tramways began to standardise on scarlet. On the LUT at least standardization took some time – No. 218, stored at Fulwell, kept its blue livery until the last in 1931.

Originally, cars had seat cushions and curtains in the saloon. Scarlet cars had red cushions with a blue or black pattern. White cars had cushions and curtains of golden russet. Blue cars had blue cushions with a red flower design. Both cushions and curtains were later discarded as they were found to harbour vermin, and the saloon seating then became woven rattan. Seats on the top deck were of the throw-over 'garden' type with slats.

Normally, drivers wore caps with raffia tops, but in bad weather those driving uncanopied cars, at least, donned a sou'-wester, as well as mackintosh, cape and leggings. Thus attired, they resembled a lifeboat coxswain in appearance. The nautical effect was enhanced by the brass of the brake handle, which spun round fast when the driver released a ratchet with his toe, and the resemblance, noted by one journalist, of the open-air reversed staircases of the uncanopied cars to the approach to the fighting top of a warship!

On cars Nos 1–300 the trolley booms turned on one side of the car only. On those later fitted with top covers the conductor had to adopt the dangerous practice of 'walking' the trolley boom round on the road at termini.

Cars which retained their open tops subsequently received hoops at each end of the top deck. The hoop was a rectangular frame designed to prevent a dewired boom from causing injury to passengers riding outside.

Boards replaced roller blind destination indicators. Route numbers were shown on black enamelled plates with white figures, fixed above the indicator box, with hooded lights to illuminate them. Type T cars later

Appendix

reverted to the use of blinds and carried a second box showing the route number.

The first cars were painted with the names of principal places served in large serif letters. This practice necessarily ceased as the system expanded and the company then adopted reversible side boards bearing the names of places served in small capitals.

Some cars of the 1–300 batches were later fitted with a white board in front of the half-landing which showed seven of the places served.

At first, cars bore no advertisements. Later, advertisements of both metal plate and paper types were carried. Typical advertisements in the earlier years were: Lipton's Tea, Zebra Gate Polish, Royal Stout, Borwick's Baking Powder and – on the same car! – Epps's Cocoa and Van Houten's Cocoa. Later additions included Glaxo, Maple & Co., Mann Crossman Beers and Black & White Whisky. Sometimes cars would carry advertisements for theatrical performances. Inside the cars were small transparent advertisements. Among them, as Mr R. E. Tustin has nostalgically recalled, were: Cats White, Homes Bright – Gospo and Hall's Wine.

As the ornate original wrought ironwork on the upper deck of open-top cars deteriorated it was replaced by double-width decency panels.

Appendix IV

Rules and Regulations

The Board of Trade regulations dated December 7, 1906, relating to the LUT show many speed restrictions classified under maxima of 14, 12, 8, 6 and 4 m.p.h. One of the 12 m.p.h. limits was between Kingston Bridge approach and the 'Greyhound', Hampton Court. Another was between Kingston Hill terminus and Alexandra Terrace. [By a regulation of September 1917, after an accident, this restriction was tightened to 8 m.p.h. and compulsory halts were enjoined before entering the facing points to the two Kingston Hill loops.]

There was an 8 m.p.h. limit through Uxbridge, in parts of Acton and Hounslow, in King Street, Twickenham, and through the main streets of Kingston. Restrictions to 6 m.p.h. applied through most of Brentford, through interlacing track in Ealing, in part of Hammersmith, through Teddington, over Kingston Bridge and elsewhere. There was a 4 m.p.h. limit through all facing points, round the London Road/Twickenham Road curve, on other sections in Twickenham, round the curve at Stanley Road junction, round the curve at St Albans Church, Teddington, and round the following curves, as well as round all curves of 66-foot radius or less:

King Street/Studland Street; Studland Street/Glenthorne Road; Paddenswick Road/Goldhawk Road; Goldhawk Road/Askew Road; Askew Road/Uxbridge Road; Boston Road/Hanwell Broadway; London Road/Richmond Road (Kingston)

There were many compulsory stops, particularly in Hammersmith, Kingston, Surbiton and Twickenham.

Additional BOT regulations of 1907, after the opening of the Malden–Wimbledon–Tooting extension, included other speed maxima, including:

16 m.p.h. in Plough Lane, Wimbledon

Appendix

14 m.p.h. in Wimbledon Broadway, through Merton and Colliers Wood and in almost all Haydons Road
12 m.p.h. in Wimbledon Hill Road and between Latimer Road and Merton High Street
8 m.p.h. over Wimbledon Railway bridge
6 m.p.h. under all low bridges
4 m.p.h. round all curves in Wimbledon

By contrast, Ministry of Transport Regulations and Bye-Laws of October 13, 1927, allowed a maximum speed of 20 m.p.h. across Ealing Common, between Ealing Town Hall and Hanwell Broadway, in parts of Southall and Boston Road and over part of the Busch Corner–Twickenham section. Elsewhere the maximum was 16 m.p.h., except for 12 m.p.h. in parts of Kingston and Wimbledon and many other places. There were certain 8 and 4 m.p.h. minima and 42 compulsory stops.

Motormen had to sound the gong when passing a busy side street and when passing another tramcar which was stationary, or a bus. They had to use it sparingly when passing churches or chapels during hours of divine service. Three gong rings was the call for an inspector and for assistance when passing a depot.

Single-track sections were governed by electric lights in boxes mounted on the nearest left-hand pole. The lights were actuated by contactor switches in the trolley wires. No light showing indicated line clear. A green light signified a car proceeding ahead on the single line and a red light a car approaching on the single line.

In later years, when fog prevailed, a 12-minute srevice was operated by five cars between Southall (Beresford Road) and Uxbridge, crossing on the loops at Beresford Road, Yeading Lane, Hayes Post Office, Heath Boys School and Greenway.

The LUT used both Turner and Collins type point controllers. Motormen operated the Turner points by putting on the handbrake and placing the controller on the second or third notch when the trolley passed underneath a skate in the wire. The points automatically reset. Cars on the 'set' line coasted under the skate with the controller 'off'. The overhead frog worked automatically with the points.

With the Collins type the points remained in the direction taken by the preceding car and were automatically operated by all cars whether the controller was 'on' or 'off', because of a separate trolley wire for each route. Cars had to pass slowly under the skate to operate the points.

Appendix

There were Turner controllers at:

Youngs Corner	set for Hammersmith, operated for Shepherds Bush
Askew Arms	set for Shepherds Bush, operated for Hammersmith
Kew Bridge	set for Brentford, operated for terminus
Busch Corner	set for Hounslow, operated for Twickenham
Stanley Road	set for Hampton Court, operated for Teddington
Kingston Bridge	set for Hampton Court, operated for Hampton Wick
London Road, Kingston (1)	set for Kingston Bridge, operated for Eden Street
(2)	set for Richmond Road, operated for Norbiton
'Liverpool Arms' (Norbiton)	set for Kingston Hill, operated for Malden

Collins controllers were at:

King Street, Hammersmith	Chiswick or Acton
Goldhawk Road (1)	Shepherds Bush or Hammersmith
Goldhawk Road (2)	Chiswick or Acton
Wimbledon Hill Road	Merton or Raynes Park

The conditions of service, rules, regulations and instructions for LUT, MET and SMET motormen and conductors in 1928 show that there was still no guarantee of permanent employment. Service adjustments might mean staff reductions, although any man affected received the option of returning in order of seniority when vacancies occurred. In these circumstances, the loyalty of the staff was commendable.

On appointment and for his first six months a man might expect 10s 8d a day. After two years' service he would receive 12s 2d. Forty-eight hours pay weekly was guaranteed for six days' work.

New entrants to the traffic department were accepted for conductor's duties only. Suitable conductors were chosen for training as motormen.

Uniforms had to be kept neat and clean. Tunics had to be worn buttoned. White cap covers were issued for use between May 1st and September 30th.

When on duty conductors were forbidden to sit inside or outside the car. They were responsible for the car's tidy appearance. They had to warn passengers leaving cars of any obstruction or opening in the road, and of vehicles approaching, and caution them against alighting before the car stopped. 'In spite of any protests, conductors will prevent ladies and children from alighting before the car has stopped.'

When a car was ascending a steep incline, the conductor was to refrain as far as possible from taking fares. He was to stand on the rear platform, ready if necessary to sand the track and apply the rear brake.

Appendix

In general, the LUT used side standards with span wire. Centre standards were originally used in Ealing and Southall but later gave place to side standards (as stated in main text). The standards were topped by a spiked finial and had crossbars with scrollwork. An ornate cast-iron casing enclosed the base. A considerable number of standards remain in situ for use as lighting standards – one, at least, known to the author, in Francis Grove, Wimbledon, still has its cast-iron casing.

Standards were painted leaf-green and bore numbers in white stencil. Painting and repainting were done by contract, per post.

At fare stages the pole bore a white disc indicating the stage number. 'Stop' signs were of white enamelled iron with a red target and black letters bearing the legend: 'Electric Cars Stop Here'. Request stop signs were in green with white letters. Later, a buff plate with red circle and bar, and the words TRAMWAYS and TRAM STOP were used. On sharp corners, poles displayed a notice with the words: 'Electric Cars Must Not Pass Each Other On This Curve'.

Bibliography

PERIODICALS

Acton and Chiswick Gazette
Chiswick Times
Daily Telegraph
Electrical Review
Electrical Times
Evening News
Irish Railway Record Society Journal
Light Railway and Tramway Journal, later Electric Railway, Bus and Tram Journal
Middlesex County Times
Modern Tramway
Modern Transport
Omnibus Magazine

Railway Gazette
Railway Magazine
Railway World, later Tramway and Railway World
Richmond and Twickenham Times
Surrey Comet
Tatler
Thames Valley Times
The Times
Tramway Review
West London Advertiser
West London Observer
Wimbledon Boro' News

OTHER PUBLICATIONS AND SOURCES

A History of London Transport, Vol. 1, T. C. Barker and Michael Robbins (Allen & Unwin)
By Tube and Electric Car to Hampton Court, W. T. Pike & Co. for LUT
Electric Traction: London's Tubes, Trams and Trains, J. C. Robinson, Society of Arts
Electric Tramways for Richmond, W. T. Pike & Co. for LUT
London General: The Story of the London Bus, London Transport
London's Trolleybuses: A Fleet History, The PSV Circle and the Omnibus Society
Royal Commission on London Traffic, 1903-5
Souvenir of the Inauguration of the London United Electric Tramways, 1901
The Felthams: The Story of the Union Construction Company, Dryhurst Publications
The Golden Age of Tramways, C. F. Klapper, Routledge & Kegan Paul
100 Years of the District, Charles E. Lee, London Transport
The London United Tramways: a Short History, B. Connelly, Tramway and Light Railway Society
Tramway Memories (especially contribution Robinson's Empire by R. E. Tustin) ed. J. Joyce, Ian Allan
Tramways of the World, address by Sir J. C. Robinson to the Tramways and Light Railways Association, 1908
To Uxbridge from the City by Train, Tube and Car via Ealing or Harrow, W. T. Pike & Co. for LUT
LUT Minutes and Reports

Index

accidents, 64, 103, 111, 148, 150, 151, 157, 160
Acton, 20, 21, 22, 23, 24, 27, 35, 36, 38, 40, 41, 46, 47, 48, 50, 51, 60, 68, 76, 103, 127, 146, 151, 161, 163, 178
Acton Depot, 27, 36, 103
Acworth, Sir W. M., 141, 142, 143, 151, 162
Arts, Society of, 90
Ashfield, Baron, of Southwell (formerly Stanley, Albert Henry), 127, 128, 129, 135, 141, 144, 147, 152, 158–9, 166, 167, 176, 177
Asquith, Lord Oxford and, 72
Associated Equipment Co. Ltd, 157, 167, 171, 177

Baber Bridge, 44, 49, 51, 73, 83, 84, 99, 119
Baker, Sir Benjamin, 63
Balfour, Rt Hon. Arthur, 63
Barber, A. L., 141
Barnes, 21, 59, 68, 70, 71, 79, 80, 81, 82, 83, 85, 120
Bessbrook & Newry Tramway, 24
Billinton, George, 20
Birkenhead, 18, 29
Blain, H. E., 129, 132
Board of Trade, HM., 19, 46, 60, 78, 82, 92, 98, 100, 104, 110, 113, 116, 119, 123, 124, 128, 147, 152
Bradford Corporation Tramways, 137, 157
Brentford, 21, 41, 44, 60, 65, 68, 72, 98, 99, 100, 103, 120, 139, 143, 147, 148, 150, 151, 152, 165, 178
Brentford & District Tramways, 23
Brentford, Isleworth & Twickenham Tramways, 21, 23
Bristol Tramways Company and Bristol Tramways & Carriage Company, 27, 32, 43, 52, 59, 141, 147
British Electric Traction, 53, 72, 78, 80, 97, 141, 145
British Thomson-Houston, 43, 46, 52, 141, 176
Brompton & Piccadilly Circus Railway, 87
Bruce, J. K., 135
Bruce, Lewis, 110, 111
Buckinghamshire, 99
Bull, Sir William J., 63, 74, 106, 133

buses, horse, 21, 51, 52, 67, 76, 112, 115
buses, motor, 75–6, 111, 115, 118, 125–6, 129, 142, 158–9, 162, 164, 167

Cassel, Sir Ernest, 63
Cater Scott, Charles James, 105, 106, 111, 121, 125, 126, 127, 129, 130, 138, 139, 141
Cecil, Lord Robert, 70, 71
Central London Railway, 57, 63, 67, 84, 90, 168
Chiswick, 17, 22, 24, 25, 44, 46, 48, 57, 58, 63, 66, 74, 76, 96, 106, 110, 119, 138, 147, 151, 152, 154, 155, 156
Chiswick Depot and Power Station, 24, 48, 57, 58, 63, 74, 96, 106, 110, 119, 147, 152, 154, 155, 156
City & North East Suburban Railway, 85
City & South London Railway, 90
City of London & Southwark Subway, 30
Clough, Smith & Co. Ltd, 167, 171, 173
Colliers Wood, 79, 80, 117
Colnbrook, 49, 77, 84, 99
conduit operation, 44, 46, 50, 60, 91, 99, 105
Cork Tramways, 29
Cramp, Charles Courtney, 21
Cranford, 22, 49, 68, 71, 73, 84, 99
Crompton, Colonel R.E., 75
Croydon, 47, 78, 79, 80, 157
Curtis, John, 18

Datchet, 77, 84
Devonshire, Sir James, 141, 152
Dittons, The, 54, 79, 110, 112, 113, 173
Drake & Gorham Electric Power & Traction (Pioneer) Syndicate, 44, 45, 53
Dublin Southern District Tramways, 32, 33–4, 52, 123, 141

Ealing, 20, 23, 37, 38, 40, 41, 44, 46, 47, 50, 63, 64, 66, 67, 68, 99, 101, 104, 113, 119, 120, 121, 137, 138
Edwards, E. H., 172
Electro-Magnetic Brake Co., 168
English Electric Co. Ltd, 167, 171, 176, 177
Eppelsheimer, E. S., 30, 31
Esher, 54, 79, 112, 113
Everard, Edward, 59

Index

Fay, Sir Sam, 45
Fell, A. L. Coventry, 126
Fisher & Parrish, 29
Fox, Sir Douglas, 82
Franco-British Exhibition, 122
Fulwell Depot and Substation, 92, 95–6, 110, 117, 154, 156, 158, 165, 167, 172, 174, 175

Garrick Villa, 88, 96, 103, 106, 119, 121
Giants Causeway, Portrush & Bush Valley Railway and Tramway, 24
Gibb, Sir George, 172
Gloucester tramways, 32
Godfray, Hugh Charles, 59
Great Northerl & Strand Railway, 87
Great West Road, 143, 162
Great Western Railway, 39, 77, 84, 101
Green, Sir William Curtis, 59
Greenwood & Batley, 53

Hallidie A. S., 30
Ham, 118
Hammersmith, 17, 20, 22, 23, 24, 25, 27, 36, 44, 45, 46, 50, 54, 55, 60, 67, 68, 70, 71, 80, 81, 83, 85, 99, 100, 102, 103, 118, 119, 120, 124, 137, 141, 145, 147, 148, 151, 152, 155, 165, 178
Hampton (including Hampton Hill), 44, 45, 49, 58, 68, 73, 82, 88, 95, 97, 103, 124, 155, 162, 165
Hampton Court, 44, 45, 51, 54, 69, 71, 88, 96, 97, 102, 122, 152, 163, 173, 174, 177, 178
Hampton Wick, 44, 45, 53, 54, 82, 92, 95, 110, 111, 119, 130, 131
Hanwell, 20, 23, 24, 38, 40, 41, 42, 44, 47, 50, 51, 58, 103, 110, 119, 120, 147, 151, 154, 155, 164, 165, 178
Hanwell Depot and Substation, 58, 103, 110, 119
Hanworth, 68, 73, 105
Hayes, 101, 102, 151, 155
Henley, W. T. Telegraph Works Co., 109
Heston, 49, 104
Heston-Isleworth Urban District Council, 23, 40, 41, 71, 83, 90, 119–20, 127, 131, 138, 141
Hillingdon, 39, 103, 110
Hillingdon Depot and Substation, 103, 110
Hook, 54, 113
Hounslow, 21, 22, 23, 38, 40, 41, 44, 45, 49, 51, 60, 65, 68, 71, 72, 73, 77, 84, 104, 138, 147, 155, 161, 178
Hounslow Depot and Substation, 58, 65, 103, 110, 161

Hounslow Heath, 21, 44, 51, 84, 154, 155 (*see also under* Baber Bridge)
Hurst, Nelson & Co., 52, 66

Immisch Electric Launch Company, 97
Imperial Tramways Company, 28, 32, 42, 43, 52, 59, 91, 121
Isleworth, 21, 23, 49, 65, 68, 90, 104, 138

Jekyll, Sir Herbert, 129

Keith House (Porchester Gate), 106
Kempton Park, 105, 155
Kensal Green Cemetery, 135–6
Kensington, 23, 24, 59, 67, 68, 85
Kew Bridge, 17, 21, 22, 23, 39, 40, 46, 57, 70, 83, 99, 120, 121, 123, 139
Kew Observatory, 44, 57, 60, 90
Kew, Richmond & Kingston-on-Thames Tramways, 52
Kincaid, Waller & Manville, 54
Kingston-on-Thames, 39, 44, 45, 52, 53, 54, 55, 56, 57, 59, 69, 71, 78, 79, 80, 81, 103, 108, 109, 110, 112, 113, 124, 130, 143, 148, 151, 153, 155, 156, 160, 162, 165, 167, 171, 173, 176
Knapp, Zac Ellis, 129, 135
Krauss, August, 26

Langley, 84
Leatherhead, 80
Leeds, 33, 37, 154, 178
Light Railway Commissioners, 38, 44, 45, 48, 49, 53, 71, 72, 84
Light Railways: Kew Bridge–Kingston 39; Acton–Hanwell, 44; Hounslow Barracks–Richmond Bridge, 45; Middlesex County Council, 49, 50; Hanwell–Uxbridge, 50, 100, 105; Clapham–Beverley Brook, 53; Southall–Hounslow, 72, 73; Hounslow–Hanworth–Sunbury, 73; Twickenham–Hanworth, 73; Hampton–Sunbury, 73; Chiswick, 76; Hounslow, Slough & Datchet, 77; Tooting Junction–Croydon, 78; Cranford–Maidenhead, 84; Farnham–Haslemere, 91; Tottenham–Walthamstow, 161
Lineff conduit system, 24, 58
Lipton, Sir Thomas, 106
Littler, Sir Ralph, 40, 49, 50
Liverpool, 18, 19, 143, 154
London & South Western Railway, 45, 48–9, 51, 54, 57, 65, 72, 79, 82, 83, 91, 108, 152

Index

London & Suburban Omnibus Co., 111
London & Suburban Traction Company, 141, 145, 147, 158-9
London, Brighton & South Coast Ry, 126
London County Council, 36, 44, 46, 50, 53, 55, 78, 80, 81, 82, 83, 91, 110, 117, 118, 120, 121, 126, 137, 141, 142, 145, 147, 148, 154, 162, 163, 165, 166, 174, 177
London Electric Railway, 128, 141, 172
London General Omnibus Company, 17, 18, 21, 67, 126, 137, 140, 141, 159, 162
London Passenger Transport Board, 175, 177, 178
London Traffic Act, 164
London Traffic, Select Committee on, 142
London United Electric Railway, 85-7
London United Tramways: registration, 26; initial capital, 26; signatories, 27; gains first Act, 27, 35; build Acton Depot, 27; new tramcars, 28, 36; uniforms, 28; new horses, 28, 36; new rails, 36; 1898 Bill, 38; Hounslow extension, 38; power stations, 38; Kew—Kingston Light Railway, 38-40; 1898 Act, 40; Imperial Tramways holding in, 43; extensions, 44; gains concession for Acton-Hanwell line, 44; seeks Light Railway Order for extensions, 44-5

LSWR fears traffic loss, 45; 1899 Act, 46; overhead in Hammersmith sanctioned, 46; terms of agreement with Ealing Council, 47; first electric tramcars, 52; 1899 returns, 52; agreement with Richmond, 60; Kew Observatory deadlock settled, 60; electric services begin, 60; electric tramcars praised, 60; design of overhead praised, 60; passengers carried, 1901, 1902, 61, 62, 69, 70; complaints, 61, 66, 67; formal inauguration, 62-6; Hounslow extension opened, 65; 1901 Bill, 67-8; payments to local authorities, 69; traffic prospects, 70

1901 Act, 73; coming-of-age party for Clifton Robinson, junior, 74; Clifton Robinson, junior, as traffic superintendent, 75; staff conditions, outings, uniforms, 75; Chiswick schemes, 76; assents to Wimbledon terms, 79; projected lines, October 1901, 79; agreement with BET, 80; 1902 Bill, 80; Hammersmith Bridge agreement with LCC, 81; threaten arbitration over Teddington, 82; seek agreement with Richmond, 83; 1902 Act, 83; proposed Maidenhead Light Railway, 1902, 84; lines opened, 1902, 84; share allocation, 84

LUER Bill, 85; Speyers gain control, 87; through bookings, 88; acquire Garrick Villa, 88; capital increase, 90; Twickenham lines inspected, 93; Hampton Court loop opened, 96; circular tours, 96; East End passengers, 97; workmen's cars, 97; labour relations, 97; motormen fined for speeding, 98; 1903 Act, 98; 1904 Bill, 99; rapacity of Brentford Council, 99; Slough demands conduit, 99; Hammersmith tramway subway proposed, 99, 100, 103; 1904 Act, 100; 1903 traffic, 100; 1903 results, 100

work begun on Uxbridge line, 100; trial runs over Uxbridge and Askew Road lines, 100; accident at Hillingdon, 103; Cater Scott appointed chairman, 105; Board congratulates Robinson on knighthood, 105; staff honour Robinson, 106; 1905 results, 107; 1905 and 1906 Acts, 107; contracts for Surrey lines, 108; Kingston lines inaugurated, 110-11; complaints of noise, 113, 119; Malden-Raynes Park service inaugurated, 116; Surrey depot site sought, 117, 118, 126; Clifton Robinson, junior, resigns, 118

1907 Bill and Act, 119-20; through running with LCC, 120-1; 1908 Bill and Act, 120-1; 1908 results and dividend, 121; Hammersmith arbitration, 124; speed limits, 124; 1908 Report and Accounts, 125; bus competition, 125; hit by building recession, 126; reciprocal running with LCC, 126; Hammersmith purchase, 126; 1909 Bill, 126; interchange at Acton, 127; Robinson resigns as Managing Director and engineer, 127; Stanley appointed managing director, 127; Board recognizes Robinson's services, 128; Robinson on condition of undertaking, 128

1909 receipts, 130; car overhaul and repair, 131; welding of joints, 131; covering of open-top tramcars, 131; staff at Robinson's funeral, 135; complaints about track, 137; car overhaul in 1911, 137; ticket interchange, 137; powers as to Chiswick UDC trolleybuses, 138; 1911 revenue and traffic, 138-9; staff conditions, 1911, 139; track, state of, in 1911, 139; top-covering of cars, 139; Kew Bridge link,

Index

139; Kew Road line abandonment, 139–40; and London & Suburban Traction Co., 141; BET control, 141 track relaying, 141; appointments, 141; 1913 receipts, 142; shareholdings, 142; trials for through running at Tooting, 142; seeks power to operate trailers, 143; 1914 Act, 143; effects of war, 145; offices changed to St James's Park, 145; 1914–15 results, 145; women conductors, 145; link with MET at Acton, 146; track reconstruction, 147; 1916 results, 147; Hammersmith arrangements, 147, 148; receiver appointed, 148; 1917 results, 148; 1918 Bill, 148; dividends and recapitalization, 149; Spencer appointed general manager and engineer, 149

fare revision, 151; accidents, 1918–19, 150–2; reduction of capital, 151; track renewals, 151; 1918 results, 151; wages and materials, 1918, 151; changeover to Lots Road supply, 152; 1919 results, 152; speed increase sought, 152; fares increase, 152; effect of LSWR electrification, 152; 1920, 1921 results, 153; one-man car operation, 153: LCC takes over Tooting-Wimbledon operation, 154; LCC through running to Kew Bridge, 154; Richmond Bridge route considered for trolleybuses, 154; Middlesex CC proposes Uxbridge Road sleeper track, 154–5, 161; Hounslow Heath section abandoned, 154; track relaying, 1922, 155

trolleybus trials on Summerstown route, 157–8; 1924 Bill for trolleybuses on Richmond Bridge route, 158–9; 1922, 1923 results, 159; Richmond Bridge route abandoned, 160; one-man operation extended, 160; 1924 results, 161; modernized tramcars, 161; bus competition, 162; LCC through working to Hampton Court, 162; 1925 results, 162; 1926 results, 163; 1927, 1928 results, 163; joint Acton-Putney service with LCC, 163; proposed reservation on Uxbridge Road, 164; Surrey lines problem, 165

Bill for trolleybus operation, 165, 167; Feltham (UCC) cars ordered, 167, 168; 1929 results, 167; Act for trolleybus operation, 167; 1930 results, 168; Feltham cars described, 170; trolleybus inauguration, 172; trolleybus extensions, 173; completion of trolleybus conversion programme, 174; trolleybus operation statistics, 175; Tolworth extension, 176–7; takeover by LPTB, 177; LPTB converts all ex-LUT to trolley-bus, 178

Longford, 72
Los Angeles, 31
Lots Road Power Station, 86, 109, 113, 148, 152
LUT Athletic Club, 156

Metropolitan Association of Electric Tramway Managers, 129, 132
Metropolitan District Electric Traction Co. Ltd, 72, 77, 84, 87 (*see also under* Underground Electric Railways of London Ltd)
Metropolitan District Railway, 23, 49, 72, 86, 99, 105, 116, 128, 138, 172
Metropolitan Electric Tramways Ltd, 72, 127, 141, 143, 146, 153, 161, 164, 167, 168, 169, 170, 174, 177
Metropolitan Railway, 20, 101, 118
Metropolitan Street Tramways, 20
Metropolitan Tramways & Omnibus Co. Ltd, 72
Maidenhead, 84; Bridge, 84, 99
Malden, 53, 54, 57, 79, 80, 81, 83, 84, 108, 112, 116, 130, 162, 165, 170, 176
Marble Arch, 67, 80, 85
Merton, 78, 79, 81, 83, 108, 116, 117, 124, 154, 155, 157, 158, 165
Middlesbrough, 32, 42, 43, 135
Middlesbrough, Stockton & Thornaby Electric Tramways, 42, 43, 74
Middlesex County Council, 39, 40, 45, 49, 50, 51, 56, 69, 71, 72, 76, 107, 123, 124, 127, 137, 138, 143, 154, 161, 164, 167
Milnes, G. F. & Co., 26, 27, 33, 34, 35, 42, 66
Mitcham, 78, 79, 80, 81, 83, 157
Molesey, 54, 62, 69, 113
Morden, 79, 124, 165
Morgan, John Pierpont, 63, 85, 86, 87
Mowlem, John, & Co. Ltd, 108
Municipal Tramways Association, 150

New Electric Traction Company, 46
New York, 18, 29, 133
North East London Railway, 85
North Metropolitan Tramways, 20, 142

overhead equipment, 24, 32, 36, 37, 41, 46, 47, 48, 50, 60, 63, 91, 109, 120, 142, 146, 157–8, 167, 171, 177

Index

Paddington, 69
Pearson, Janet, 156
Perks, Sir Robert, 86, 87
Petersham, 79, 118
Piccadilly & City Railway, 85
Piccadilly, City & North East London Railway, 85, 86
Pick, Frank, 172, 176
Pimlico, Peckham & Greenwich Tramway, 20
Preece, Sir William, 63
Putney, 54, 59, 60, 67, 68, 148, 163
Putney Vale, 53

Ranelagh Club, 70
Raynes Park, 78, 79, 80, 83, 116, 118
Reading, 32, 43
Reckenzaun, Anthony, 22
Richmond, 17, 21, 22, 26, 27, 30, 35, 39, 40, 44, 45, 46, 57, 59, 60, 68, 69, 70, 71, 79, 80, 81, 82, 83, 84, 102, 103, 111, 118, 120, 138, 139, 140, 155, 158, 160
Richmond Depot, 26, 140, 155
Richmond Electric Light & Power Company, 120
Richmond, Virginia, 24, 32
Robinson, Sir James Clifton, 27, 44, 50, 80; birth and parentage, 29; fascinated by Train's tramcar, 29; goes to New York as protegé of Train, 29; returns to England, 29; with Fisher and Parrish, 29; becomes general manager at Cork, 29; marries Mary Edith Martin, 29; becomes general manager at Bristol, 30; son, Clifton, born, 30; becomes general manager at Edinburgh, 30; manages Hallidie Corporation, 30; in Los Angeles, 31, 32; goes to San Francisco, 32; becomes managing director of Imperial Tramways Co., 32; reports on electrification of Bristol tramways, 33
becomes managing director of Dublin Southern District tramways and electrifies system, 33-4; opens tram campaign in Ealing, 37; as newspaper proprietor, 37; takes house in Goldhawk Road, 37; seizes on Light Railway Act, 38; bows to Richmond opposition, 40; wins over Ealing, 47; 'propaganda machine', 47; forecasts LUT traffic, 51, 58, 70; electric tramcar design, 52; and Surrey schemes, 53-4; and Kingston lines, 55; settles Kew Observatory deadlock, 57, 60; shows Chiswick depot to councillors, 58; determines on Richmond, 59; on punitive local demands, 62

on 'audacity' of Middlesex Council, 70; at son's coming-of-age party, 74; on buses and tramcars, 76, 115; on air transport, 76, 132; on tube railways, 76; accepts Wimbledon terms, 79; at examination of 1902 Bill, 81; contemplates tube railway, 84; on London United Electric Railway plans, 85; described by *Chiswick Times*, 91; described by *Middlesex County Times*, 93; on Hampton Court loop trial run, 95; at Hampton Court loop opening, 96; benevolent autocrat, 97; on LCC's attitude to Hammersmith subway proposal, 100; biographical note, 102
evidence before Royal Commission on London Traffic, 104; knighted 105; distinctions and offices, 106; recreations, 106; friendships, 106; honoured by staff, 106; praises staff, 107, 112; at inauguration of Kingston lines, 110-11; sustains injury, 111; writes to Board of Trade about noise, 113; *Railway Magazine* interviews, 115-16; announces son's resignation, 118; salary 120; proposed as Tees-side parliamentary candidate, 120; world tour, 120
on tramways in Japan and United States, 121; on double-versus single deck tramcars, 121; on tramcar speed, 121; addresses Tramways & Light Railways Association, 122; refutes allegations by Twickenham, 124; resigns as managing director and engineer, 127; accepts commission from Speyer Bros, 128; on restrictions on British tramway enterprise, 128; leaves for Far East, 128; on good condition of LUT, 128; gives dinner to Metropolitan Assoc. of Electric Tramway Managers, 128-9; refers appreciatively to A. H. Stanley, 128-9; returns from Far East, 132
receives commission to visit Newfoundland, 132; entertained by Metropolitan Assoc. of Electric Tramway Managers, 132; recalls G. F. Train, 132; assignment in Canada, 132; goes to New York, 133; death, 133; obituary notices, 133-5; funeral, 135; grave, 136; Will, 136; 'wraith of', 160
Robinson, Clifton, Junior, 30, 64, 74, 75, 92, 93, 96, 102, 106, 118, 128
Robinson, Lady Mary, 29, 93, 103, 106, 119, 120, 121, 128, 133, 160
Rogers, Reuben Cramp, 74, 118, 129
Rothschild, Lord, 63

238

Index

Royal Society, 44, 57

San Francisco, 30, 32
Seal, Mrs E., 145
Shepherds Bush, 17, 20, 21, 22, 23, 24, 25, 28, 35, 37, 41, 46, 57, 63, 64, 67, 80, 85, 125, 156, 170, 174
Shepherds Bush Depot, 25, 35
Siemens, Alexander, 41
Siemens, Werner von, 24
Slough, 77, 84, 99
Southall, 20, 23, 39, 49, 50, 68, 72, 73, 76, 100, 101, 104, 105, 127, 138, 154, 155
Southall Ealing & Shepherds Bush Tram-Railway Co., 20, 21, 22
Southall Hounslow & Twickenham Railless Traction Company, 138
South Metropolitan Electric Tramways & Lighting Co. Ltd, 78, 141, 157
Spencer, Christopher John: birth and early career, 150; advises Chiswick UDC on trolleybuses, 137; appointed general manager of LUT and MET, 149; launches track renewal programme, 157; studies United States tramways, 153; original views of, 152; tries one-man cars on LUT, 153; Spencer-Dawson brake, 154; considers Richmond Bridge line for trolleybuses, 154; extends one-man operation, 160-1; modernizes tramcars, 161; as motorist, 162; gives evidence before Royal Commission on Transport, 164; on trolleybus design progress, 165; on LUT trolleybus conversion scheme, 167; on anti-tram fallacies, 172; at Kingston area trolleybus inauguration, 173; on case for trolleybus conversion, 174; resigns from LPTB, 178
Speyer Bros, 87, 128
Speyer, Sir Edgar, 87, 112, 127, 129
Sprague, Frank J., 24, 32
staff, 34, 60, 61, 74-5, 98, 103, 107, 119, 129, 139, 145, 146, 155, 156, 161
Staines, 49, 99, 119
Stanley, Albert Henry – *see under* Ashfield, Baron, of Southwell
Steep Grade Tramways & Works Co. Ltd, 30
Summerstown, 78, 79, 80, 83, 117, 154, 157, 165, 168
Sunbury, 68, 72, 73
Surbiton, 52, 53, 54, 59, 71, 108, 110, 112, 113, 130, 143, 165, 176
Surrey, 44, 45, 53, 54, 62, 69, 79, 82, 84, 113, 117, 123, 124, 143

Surrey County Council, 45, 53, 62, 69, 82, 84, 113, 117, 123, 143
Sutton, 79, 80
Syon House, 41
Szlumper, Sir James, 40, 106, 111, 112, 135

Teddington, 44, 47, 48, 49, 61, 68, 74, 82, 89, 91, 92, 95, 104, 124, 131, 165, 167, 172, 173, 176
Thames Ditton, 52, 54, 113
Thomas, Sir T. E., 168, 178
Tolworth, 54, 79, 110, 112, 153, 167, 173, 176
Tooting, 46, 55, 78, 79, 80, 83, 112, 116, 117, 126
tracklaying and trackwork, 18, 42, 46, 47, 48, 51, 54, 60, 79, 82, 89, 92, 99, 103, 108, 109, 113, 116, 131, 137, 139, 151, 155
Traffic, Royal Commission on London, 87, 104, 129
Train, George Francis, 18, 19, 29, 80, 106, 132
tramcars, electric, LUT: first, 48, 52; testing of, 60; design praised, 60; criticism of, 66; allocation of, 102; for Kingston services, 110; Kingston area, exchange of, 113; overhaul of, 1906 and 1907; 121; dead mileage, 126; overhaul and repair of, 1910 and 1911, 131, 137; top covering of, 139; size of LUT fleet, 142; trailers, 143; one-man operated, 153, 160-1; modernization, 161; Feltham type, 167-8, 169, 170, 175; taken over by LPTB, 177
tramcars, horse, LUT, 27, 28, 35, 36, 46, 48, 57, 58, 139, 140
Tramways Act, 1870, 19, 38, 123
Tramways & Light Railways Association, 122, 133, 135, 162, 164
Tramways, Light Railways & Transport Association, 172, 175
Tramways (MET) Omnibus Co., 141
Transport, Institute of, 130
Transport, Ministry of, 161, 162, 165, 173, 176
Transport, Royal Commission on, 164, 165
trolleybuses: 1911 schemes, 137; in Leeds and Bradford, 137, 138; Chiswick UDC Bill, 137-8; Spencer, C. J., early interest in, 150; Spencer, C. J., considers for Richmond Bridge route, 154; trackless cars/Blackburn vehicle, 154; AEC prototype, 157; LUT Bill for,

239

Index

165; LUT Act for, 167; LUT order for, 167; AEC/English Electric vehicle, 167; Karrier-Clough vehicle, 168; UCC-built vehicles, 171; LUT inauguration of, 172; LUT extends services, 173; LUT conversion completed, 174; interference with BBC programmes, 174; Ministry of Transport regulations for, 176; last LUT vehicle, 177; fleet taken over by LPTB, 177; LPTB conversion plans, 178

Twickenham, 21, 40, 44, 45, 49, 51, 59, 61, 68, 73, 74, 82, 83, 84, 89, 92, 93, 104, 119, 124, 127, 131, 140, 155, 158, 159, 160, 162, 165, 167, 171, 172, 173

Underground Group (Underground Electric Railways of London Ltd), 86, 88, 137, 141, 145, 157, 164, 166, 168, 172, 176, 177

Union Construction & Finance Co. Ltd, (UCC), 168, 170, 171

Uxbridge, 20, 38, 39, 44, 50, 100, 101, 102, 105, 151, 164, 167, 168

Verdon Smith, W. G., 110, 135, 141
Volk, Magnus, 24

Wandsworth, 53, 70, 168

West Ham Corporation Tramways, 129, 132
West London, Barnes & Richmond Tramway Company, 120
West London Tramway Company, 25
Westminster, 46
White City, 122
White, J. G. & Co., 108
White, Samuel, 59, 84, 87, 96
White, Sir George, 26, 30, 33, 59, 64, 84, 87, 135, 147
Whiteleys, 74
Whiteley, William, 106
Willesden, 68, 73
Wimbledon, 55, 78, 79, 80, 81, 83, 84, 110, 112, 116, 117, 118, 126, 131, 143, 154, 157, 165, 167, 168, 173, 174, 177
Winter's Bridge (Window's Bridge)—*see under* Dittons, The
Wimpey, G. & Co., 36
Woods–Gilbert Rail Remodelling Co. Ltd., 150
Works, Metropolitan Board of, 19, 24

Yerkes, Charles Tyson, 63, 72, 86, 88, 96, 103, 105, 168
Yorke, Lt.-Colonel H. A., 60, 101, 110, 111, 116, 117, 120, 139
Young's Corner – *see under* Hammersmith